Design of Concrete Buildings for Earthquake and Wind Forces

According to the 1997 Uniform Building Code

by David A. Fanella and Javeed A. Munshi

PORTLAND CEMENT ASSOCIATION

An organization of cement manufacturers to improve and extend the use of portland cement and concrete through market development, engineering, research, education, and public affairs work.

5420 Old Orchard Road, Skokie, Illinois 60077-1083

Printed in U.S.A.
Library of Congress catalog card number 98-67486
ISBN 0-89312-195-9

PREFACE

The primary purpose of this publication is to assist the engineer in the proper application of the seismic requirements of the 1997 edition of the Uniform Building Code (UBC). The 1997 UBC has adopted, with some modifications, the seismic detailing requirements of the 1995 edition of "Building Code Requirements for Structural Concrete (ACI 318-95) and Commentary (ACI 318R-95)." For a comprehensive description of and background information on the design and detailing provisions, the reader is referred to the commentary of "Recommended Lateral Force Requirements and Commentary of the Structural Engineers Association of California," (SEAOC Blue Book), 1996 edition and to the commentary of ACI 318-95.

Special emphasis is given to the detailing requirements for a variety of structural components in regions of low, moderate, and high seismic risk. The detailed design of three reinforced concrete buildings in high seismic zones utilizing the various structural systems recognized in the 1997 UBC is illustrated. All three buildings are analyzed and designed for the combined effects of gravity, seismic, and wind loads.

As always, every attempt has been made to be as accurate as possible in the design aids and design examples. Readers are urged to bring any errors or misprints they may find to our attention. Any suggestions for improvement would also be gratefully received.

Acknowledgement

Thanks are due to Cami Lobb of PCA staff who was responsible for the desktop publishing, including figures, layout, and design of the publication. Her help in the production of this manuscript is gratefully acknowledged.

David A. Fanella and Javeed A. Munshi
Engineering Services

Contents

1 INTRODUCTION

1.1 SCOPE

This publication is meant as a guide on the analysis and design of reinforced concrete buildings subjected to earthquake and wind forces. It outlines the basic considerations involved in earthquake-resistant design in areas of low, moderate, and high seismic activity. The discussion and illustrative examples that are contained in subsequent chapters of this publication are based on the provisions of the 1997 Uniform Building Code (UBC) (Ref. 1.1). The 1997 UBC has adopted the requirements of the 1995 ACI 318 Building Code (Ref. 1.2) for the design and detailing of structural members, with some modifications. It is assumed that the reader has a copy of the 1997 UBC when using this book.

The following sections highlight the major changes in the seismic design provisions contained in the 1997 UBC. Some of the changes reflect the lessons learned from the observed behavior of structures during the 1994 Northridge earthquake in California and the 1995 Hyogoken-Nanbu earthquake in Kobe, Japan. Comparisons between the requirements of the 1997 UBC and the 1994 UBC (Ref. 1.3) are also given. It is readily apparent that the new provisions can have a significant effect on the design and detailing of some structures, especially those in high seismic risk areas.

The conversion from allowable (working) stress design to strength design is clearly the most significant change from the 1994 UBC to the 1997 UBC. This and other changes in the earthquake design requirements are described in Sect. 1.2. The major changes in the concrete related provisions are summarized in Sect. 1.3.

Chapter 2 outlines the general requirements in the 1997 UBC for the analysis and design of buildings subjected to earthquake forces, including methods for determining the minimum design lateral forces due to the design basis ground motion. Included is a description of the parameters that form the basis of seismic design, as well as discussions on load combinations, vertical and horizontal force distribution, overturning, story drift limitations, and P-Δ effects.

A summary of the seismic detailing requirements is provided in Chapter 3. As noted above, these requirements are primarily from ACI 318-95, with some modifications. Included are the provisions for frame members subjected primarily to bending, frame members subjected to bending and axial loads, shearwalls, diaphragms, joints of frames, two-way slabs without beams, and frame members not part of the lateral force resisting system. The detailing requirements are presented based on the seismic zone.

The design of a 3-story special moment-resisting beam-column frame subjected to the combined effects of gravity, wind, and earthquake forces is illustrated in Chapter 4. Chapter 5 illustrates the design of a 10-story building frame system. The design and detailing of a 12-story dual system is contained in Chapter 6.

Throughout this publication, section numbers in the 1997 UBC are given in non-italicized type (e.g., 1629.1), while those in the 1994 UBC are italicized (e.g., *1629.1*).

1.2 Major Changes in Earthquake Design Requirements

1.2.1 Soil Profile Types

The four site coefficients (S_1 to S_4) in the 1994 UBC have been expanded to six soil profile types in the 1997 UBC (S_A to S_F; see Table 16-J). Based on studies completed in the 1970's, the site factors in the 1994 UBC are independent of the level of ground shaking. Recent data has shown that site factors are actually nonlinear and depend on both the input acceleration (zone factor) and the site soil characterization (site factor). A large amount of data that covers a wide range of ground shaking and site characteristics is available from the extensive strong ground motion instrumentation program that has been undertaken since the 1971 San Fernando earthquake. The data from the 1985 Mexico City and 1989 Loma Prieta earthquakes have, in particular, contributed to the new site categorization procedure and coefficients as well.

The new soil profile types are based on soil characteristics for the top 100 ft (30 m) of the soil. The properties used to identify the soil profile type are the shear wave velocity, standard penetration test, and undrained shear strength. A procedure for determining the soil profile type for a given site is contained in 1636. Type S_F soil profiles that require site-specific evaluations are given in 1629.3.1. Both the acceleration-dependent seismic coefficient (C_a) and the velocity-dependent seismic coefficient (C_v) used in determining the base shear (V) both depend on the soil profile type (see Tables 16-Q and 16-R).

1.2.2 Structural Framing Systems

In addition to the four basic framing systems (bearing wall, building frame, moment-resisting, and dual), two new structural system classifications have been introduced in the 1997 UBC: cantilevered column building systems and shearwall-frame interaction systems (see Table 16-N). A cantilevered column building system consists of a structure supported on columns that are fixed at the base and that have minimal moment capacity at the top. These columns support the gravity loads and provide resistance to the lateral loads that are essentially applied at the top of the column. Thus, failure of the columns would result in failure of the LFR system as well. This system can be used to simulate a weak story in a building.

A combination of shearwalls and frames resist the lateral forces in a shearwall-frame interaction system. Interaction between the walls and frames are considered at every level, and the lateral forces are allocated to these elements based on their relative rigidities. Since this type of system is to be used in Seismic Zones 0 and 1 only (see footnote 8 of Table 16-N), no special detailing is required for the shearwalls or ordinary moment-resisting frames.

1.2.3 Load Combinations

As noted above, the 1997 UBC seismic design provisions are based on strength-level design rather than service-level design. Thus, in accordance with this change, the load combinations have changed. Table 1-1 contains a comparison of the load combinations in the 1994 and 1997 UBC applicable to concrete structures subjected to the combined effects of dead (D), live (L), snow (S), wind (W), and earthquake (E) forces.

Table 1-1 Comparison of Load Combinations in the 1994 and 1997 UBC

1994 UBC		1997 UBC	
Eq. No.	Load Combination (*1909.2, 1921.2.7*)	Eq. No.	Load Combination (1612.2, 1909.2)
(9-1)	$1.4D + 1.7L$	(9-1)	$1.4D + 1.7L$
(9-2)	$0.75(1.4D + 1.7L + 1.7W)$	(9-2)	$0.75(1.4D + 1.7L + 1.7W)$
	$1.4(D + L + E)^*$	(9-3)	$0.9D + 1.3W$
	$0.75(1.4D + 1.7L + 1.87E)^\dagger$	(12-5)	$1.1[1.2D + 1.0E + (f_1L + f_2S)]$
(9-3)	$0.9D + 1.3W$	(12-6)	$1.1(0.9D + 1.0E)$
	$0.9D + 1.4E^*$		
	$0.9D + 1.43E^\dagger$		
	Special Load Combination (*1628.7.2*)		Special Load Combination (1612.4)
	$1.0D + 0.8L + 3(R_w/8)E$	(12-17)	$1.2D + f_1L + 1.0E_m$
	$0.85D + 3(R_w/8)E$	(12-18)	$0.9D + 1.0E_m$

R_w = response modification factor (Table 16-N, 1994 UBC)
$E = \rho E_h + E_v$ Eq. (30-1)
$E_m = \Omega_o E_h$ Eq. (30-2)
E_h = earthquake load due to the base shear (V)
E_v = earthquake effect due to vertical component of ground motion
ρ = reliability/redundancy factor
Ω_o = seismic force amplification factor (Table 16-N)
f_1, f_2 = factors defined in 1612.2.1
**Seismic Zones 3 and 4*
†Seismic Zones 2 and lower

Note that a factor of 1.1 is required in the load combinations involving earthquake effects for the design of concrete and masonry structures (see the exceptions in 1612.2.1). Further discussion on this factor is given in Sect. 2.2.

1.2.4 Earthquake Loads

In the 1997 UBC, the earthquake load (E) is a function of both the horizontal and vertical components of the ground motion, and is determined from Eq. (30-1). The parameter (E_m) is the maximum estimated earthquake force that can be developed in the structure, and is determined from Eq. (30-2). A discussion on the new parameters in these two equations is given below.

Reliability/Redundancy Factor (ρ). The purpose of this factor is to increase the effects of the earthquake loads on structures with few lateral force resisting (LFR) elements. Since these types of systems do not have multiple paths of resistance, failure of a small number of LFR elements would result in the formation of a collapse mechanism for the structure when subjected to the seismic loads. Redundancy ensures redistribution of forces as portions of the structure yield. Past earthquakes have shown that structures with a higher degree of redundancy perform better than structures with few LFR elements.

Note that ρ shall not be taken less than 1 nor greater than 1.5. A value of ρ less than or equal to 1 indicates that the structure is sufficiently redundant so that an increase in the earthquake loads (or, equivalently, an increase in strength) is not needed. On the other hand, a value of ρ greater than 1 indicates that earthquake loads must be increased to account for a lack of redundancy in the structure. Section 2.4.1 contains a comprehensive discussion on how to determine this factor for various structural systems.

Vertical Acceleration Effect (E_v). As noted above, the new provisions require that the effects of both the horizontal and vertical components of the ground motion be included when designing for seismic forces. This is reflected in Eq. (30-1) where E_h is the earthquake load due to the base shear (V) and E_v is the load effect resulting from the vertical component of the ground motion. Note that E_v is equal to an addition of $0.5C_a ID$ to the dead load effect (D). The new seismic coefficient (C_a) depends on the soil conditions, the Seismic Zone Factor (Z), and in Seismic Zone 4, the acceleration-dependent near-source factor (N_a) (see Table 16-Q). Further discussion on E_v is given in Sect. 2.4.1.

Seismic Force Amplification Factor (Ω_o). The seismic force amplification factor is utilized in the special load combinations of 1612.4 that are applicable to elements supporting discontinuous systems (see 1630.8.2 and Table 1-1). This factor is used to establish design loads for elements of the structure that are expected or required to remain elastic during the design basis ground motion (for example, columns supporting a discontinuous shearwall). In other words, elements supporting the discontinuous system must have the strength to resist the maximum load that can develop in the system. Note that Ω_o is comparable to the $3(R_w/8)$ factor which was in the 1994 UBC (*1628.7.2*).

1.2.5 Modeling Requirements

New to the 1997 UBC are guidelines for the creation of the structural model for lateral load analysis (1630.1.2). For concrete and masonry structures, cracked sections must be used in the model. For structures where deformation compatibility must be considered, the flexural and shear stiffness properties shall not exceed one-half of the gross section properties for concrete and masonry elements of the LFR system, unless a rational cracked-section analysis is performed (1633.2.4).

1.2.6 Design Base Shear

The equations for determining the design base shear (V) according to the static force procedure in the 1994 and 1997 UBC are compared in Table 1-2. Several changes are apparent. The design base shear in the 1997 UBC varies in inverse proportion to the period T, rather than $T^{2/3}$ prescribed previously. Also, the minimum design base shear limitation for Seismic Zone 4 was introduced as a result of the ground motion that was observed at sites near the fault rupture in the 1994 Northridge earthquake.

The new parameters (C_a) and (C_v) are acceleration-dependent and velocity-dependent seismic coefficients, respectively, which depend on the soil type (S_A to S_F) and the Seismic Zone Factor (Z) (see Tables 16-Q and 16-R). In Seismic Zone 4, C_a and C_v also depend on the acceleration-dependent and velocity-dependent near-source factors (N_a) and (N_v), respectively. Note that both N_a and N_v have been introduced to increase the design base shear for structures located near a known seismic source (see Tables 16-S and 16-T). The need to incorporate near-source factors became more apparent in the aftermath of the Northridge and Kobe earthquakes. These two factors alone, if applicable, could increase the magnitude of the design base shear by as much as 50 to 100 percent.

Figures 1-1, 1-2, and 1-3 show comparisons between the design base shears (V) from the 1994 and 1997 UBC for a bearing wall system, a building frame system, and a special moment-resisting frame system and a dual system, respectively, in Seismic Zone 4. In all cases, the importance factor (I) was taken as 1.00, and the soil profile type was taken as S_2 in the 1994 UBC and S_D in the 1997 UBC. In order to make the comparisons valid, the design base shears obtained from the 1994 UBC are multiplied by 1.4. The influence of the near-source factors (N_a) and (N_v) is quite apparent from these figures.

Table 1- 2 Comparison of Design Base Shear (V) Equations in the 1994 and 1997 UBC

1994 UBC	1997 UBC
$\frac{2.75ZIW}{R_w} \geq V = \frac{ZICW}{R_w} = \frac{1.25ZISW}{T^{2/3}R_w} \geq 0.075ZIW$	$\frac{2.5C_aIW}{R} \geq V = \frac{C_vIW}{RT} \geq 0.11C_aIW$ $\geq \frac{0.8ZN_vIW}{R}$ (Seismic Zone 4)
Z = seismic zone factor (Table 16-I) I = seismic importance factor (Table 16-K) C = numerical coefficient specified in *1628.2.1* W = seismic dead load (*1628.1*) R_w = response modification factor (Table 16-N) S = site coefficient (Table 16-J)	C_a = acceleration-dependent seismic coefficient (Table 16-Q) C_v = velocity-dependent seismic coefficient (Table 16-R) I = seismic importance factor (Table 16-K) W = seismic dead load (1630.1.1) R = response modification factor (Table 16-N) Z = seismic zone factor (Table 16-I) N_v = velocity-dependent near-source factor (Table 16-T)

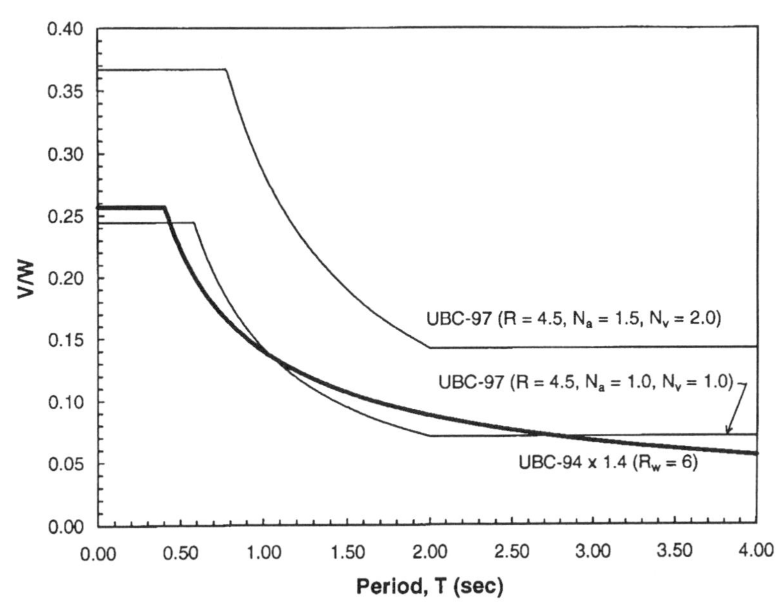

Figure 1-1 Comparison of 1994 and 1997 UBC Design Base Shears (V) for Bearing Wall Systems

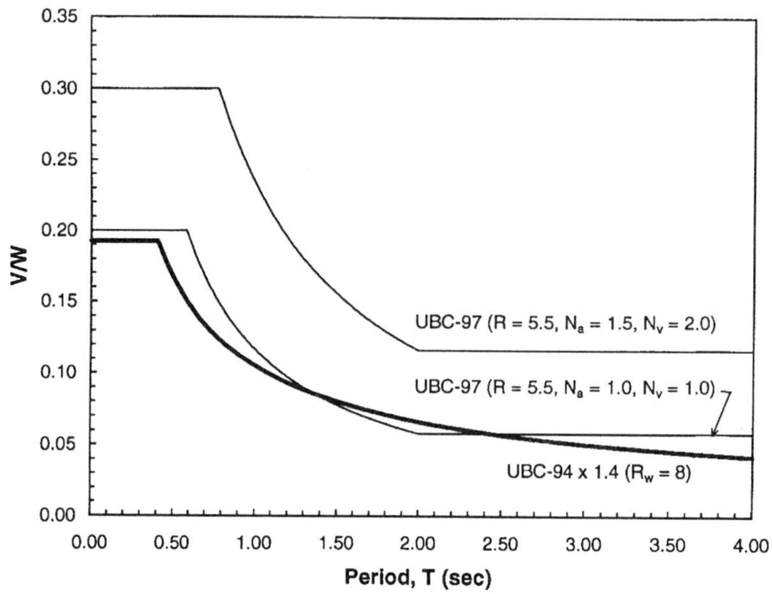

Figure 1-2 Comparison of 1994 and 1997 UBC Design Base Shears (V) for Building Frame Systems

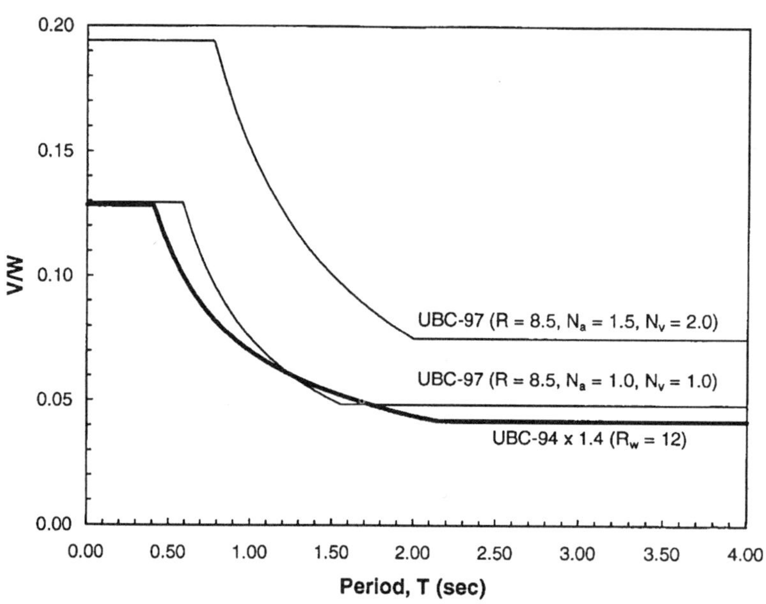

Figure 1-3 Comparison of 1994 and 1997 UBC Design Base Shears (V) for Special Moment-Resisting Frames and Dual Systems

1.2.7 Simplified Design Base Shear

New to the 1997 UBC is a simplified method for determining the design base shear (V) (1630.2.3). This method is applicable to (1) buildings that use light-frame construction of any occupancy (including single-family dwellings) not more than three stories in height, excluding basements, and (2) other buildings not more than two stories in height, excluding basements. It is important to note that when the soil profile type is not known, Type S_D shall be used in Seismic Zones 3 and 4, and Type S_E shall be used in Seismic Zones 1, 2A, and 2B. Code sections that do not apply when using the simplified method are contained in 1630.2.3.4.

1.2.8 Displacement and Drift

The displacements and drifts, which were previously determined for the service-level earthquake forces, are now determined for the strength-level earthquake forces. In the 1994 UBC, the displacements (Δ_E) are computed for the elastic structure using the code prescribed service-level forces and gross section properties. The interstory drifts based on these displacements are compared to the limiting values in *1628.8*. An estimate of the maximum inelastic displacements could be obtained by multiplying Δ_E by ($3R_w/8$). To account for the effects of cracking, the previous term could be multiplied by another factor such as 2 (i.e., assuming one-half of the gross section properties).

According to the new provisions, the calculated story drifts shall be computed using the maximum inelastic response displacement drift (Δ_M), which is an estimate of the displacement that occurs when the structure is subjected to the design basis ground motion. According to 1630.9.2, Δ_M is equal to $0.7R\Delta_S$, where Δ_S is the design level response displacement, which is the displacement that occurs when the structure is subjected to the design seismic forces. The displacement Δ_S can be obtained from either a static, elastic analysis of the LFR system using the design seismic forces from the static force procedure of 1630.2.1, or an elastic dynamic analysis in accordance with 1631. In any case, the modeling requirements of 1630.1.2 must be satisfied (i.e., cracked section properties must be used). The maximum inelastic drift (Δ_M) can also be determined directly by inelastic time history analyses of the structure. Updated limitations on the calculated story drifts using Δ_M are given in 1630.10.2.

1.2.9 Dynamic Analysis Procedures

The dynamic analysis provisions have been expanded in the 1997 UBC. In particular, 1631.6 has been added that provides explicit requirements for time-history analysis. It is important to note that when a nonlinear time-history analysis is used, a design review of the LFR system in accordance with 1631.6.3.2 by an independent engineering team is required.

1.2.10 Lateral Forces on Elements of Structures

New equations for determining the seismic force (F_p) for elements of structures, nonstructural components, and equipment are given in 1632. These new equations, based on the strength design approach, depend on the weight of the component (W_p), the component amplification factor (a_p), the seismic coefficient (C_a), the component importance factor (I_p), and the component response modification factor (R_p).

1.2.11 Deformation Compatibility

The deformation compatibility requirements of 1633.2.4 have been revised based on observations from the 1994 Northridge earthquake. All elements that are designated not part of the LFR system are required to be designed and/or detailed to maintain support of their gravity loads as they deform with the LFR system due to the seismic forces. The expected deformation is the greater of the maximum inelastic response displacement (Δ_M) or 0.0025 times the story height. As noted above, not more than one-half of the flexural and shear stiffness properties of concrete and masonry elements that are part of the LFR system must be taken for the analysis, unless a rational cracked-section analysis is performed. Also, P-Δ effects must be considered.

1.3 Major Changes in Concrete Related Provisions

The major changes outlined in the following sections apply specifically to reinforced concrete structures resisting earthquake-induced forces.

1.3.1 Welded Splices and Mechanical Connection Splices

New requirements for welded and mechanical splices have been introduced in the 1997 UBC (1921.2.6). In Seismic Zones 2, 3, and 4, welded splices on billet steel A 615 or low allow A 706 reinforcement shall not be used within an anticipated plastic hinge region, nor within a distance of one beam depth on either side of a plastic hinge region or within a joint.

Two types of mechanical connection splices are defined in 1921.2.6.1.2. In a Type 1 splice, the mechanical connectors must meet the requirements of 1912.14.3.4 and 1912.14.3.5. A Type 2 splice is a mechanical connection that must develop in tension the lesser of 95 percent of the ultimate tensile strength or 160 percent of the specified yield strength (f_y) of the bar. In Seismic Zone 1, a Type 1 splice shall be permitted at any location within a member, and in Seismic Zones 2, 3, and 4, a Type 2 splice is required within an anticipated plastic hinge region or within a distance of one beam depth on either side of the plastic hinge region or within a joint.

1.3.2 Precast Concrete

For the first time, provisions for seismic design of precast concrete structures in Seismic Zones 3 and 4 have been included (1921.2.1, 1921.2.2, and 1921.2.7). Previously, these types of structures were allowed only if it were demonstrated by experimental evidence and analysis that the system would have at least the same strength and toughness of a comparable monolithic reinforced concrete structure.

Two design alternatives for precast LFR systems are presented in the 1997 UBC (1921.2.1.6): (1) emulation of the behavior of monolithic reinforced concrete construction, and (2) utilization of the unique properties of a structural system composed of interconnected precast elements. Note that the second alternative is allowed only under the undefined structural systems provision of 1629.9.2, i.e., the response modification factor (R) must be determined by approved cyclic test data and analyses.

For emulation of the behavior of monolithic reinforced concrete construction, precast structural systems using frames must utilize either "wet" connections or "strong" connections (1921.2.2.5). A wet connection uses any of the splice methods per 1921.2.6.1 or 1921.3.2.3 to connect precast members, and uses cast-in-place concrete or grout to fill the splicing closure. Systems utilizing wet connections must comply with all of the requirements applicable to monolithic reinforced concrete construction resisting seismic forces (1921.2.2.6). Requirements for strong connections are contained in 1921.2.7. Note that requirements for precast wall systems with strong connections are not included.

Provisions for the design of structures that have a precast gravity system (consisting of structural members that are not part of the LFR system) are contained in 1921.2.1.7.

1.3.3 Shearwalls

The shearwall design provisions of 1921.6.6 that were first introduced in the 1994 UBC have been changed. In particular, 1921.6.6.2 pertaining to the effective flange width of I-, L-, C-, or T-shaped shearwalls subjected to flexural and axial loads has been modified. The second requirement of this section now states that the effective width shall be 15 percent of the total wall height for the flange in compression or 30 percent of the total wall height for the flange in tension, not to exceed the total projection of the flange. Previously, the effective width was 10 percent of the total wall height, with no distinction made between the compression and the tension flange.

The shearwall boundary zone detail requirements of 1921.6.6.4 have also been modified. The following is a summary of the changes:

1. The third condition of 1921.6.6.4 now includes $(M_u/V_u \ell_w) \leq 3$ in addition to $V_u \leq 3A_{cv}\sqrt{f_c'}$.
2. The maximum compressive strain (ε_{max}), which was equal to $3R_w$ (0.004)/8 $\leq$ 0.015 in the 1994 UBC, is 0.015 in the 1997 UBC.
3. In the 1997 UBC, the total displacement at the top of the wall (Δ_t) is equal to $\Delta_M = 0.7R\,\Delta_S$ using cracked section properties, or may be taken as $2\Delta_M$ using gross section properties. Note that Δ_S is the displacement at the top of the wall due to the strength-level design earthquake forces using the cracked section properties. Previously, Δ_t was equal to $3R_w/8$ times the elastic design displacement using cracked section properties, or $2(3R_w/8)\Delta_E$.
4. The parameter (Δ_y) was defined in the 1994 UBC as the displacement at the top of the wall corresponding to an extreme fiber compressive strain of 0.003 at the critical section. In the 1997 UBC, it is defined as the displacement at the top of the wall corresponding to yielding of the tension reinforcement at the critical section. Note, however, that the equation to determine Δ_y has not changed. Also note that M_E is now defined as the unfactored moment at the critical section when the wall is displaced Δ_E.
5. The yield curvature, which was previously given as $\phi_y = 0.003/c'_u$, is now $\phi_y = 0.003/\ell_w$.
6. In the 1997 UBC, item 1.1 of 1921.6.6.6 requires that all portions of boundary zones shall have a minimum thickness of $\ell_u/16$ where ℓ_u is the unsupported length of the wall. The 1994 UBC required a minimum thickness of $\ell_u/10$. Item 1.3 in the same section of the 1997 UBC specifies that boundary zones shall have a minimum length of 18 inches at each end of the wall or portion of wall. In the 1994 UBC, the requirement was that boundary zones shall have a minimum length of $0.1\ell_w$.
7. A new section 1921.6.6.7 is included in the 1997 UBC which pertains to welded splices and mechanical connections of longitudinal reinforcement in the boundary zone (see Sect. 1.3.1).

1.3.4 Coupling Beams

The new section 1921.6.10.1 specifies that for coupling beams with $\ell_n/d \geq 4$, the design shall conform to the general requirements of 1921.2 and the flexural member design requirements of 1921.3. Also, in 1921.6.10.2, detailing provisions are provided for coupling beams with $\ell_n/d < 4$ and reinforced with two intersecting groups of symmetrical diagonal bars. No such provisions existed in the 1994 UBC. Other new reinforcement requirements are contained in 1921.6.10.3 to 1921.6.10.5.

1.3.5 Diaphragms

The requirements for the design and detailing of diaphragms have been revised. A new requirement related to the use of mechanical connectors to transfer forces between the diaphragm and the LFR system is contained in 1921.6.12. According to the new provision, the anchorage shall be adequate to develop a force of $1.4A_sf_y$ where A_s is the cross-sectional area of the connector.

In the 1997 UBC, collectors and boundary elements in topping slabs placed over precast floors and roof elements shall have a minimum thickness of 3 in. or 6 times the diameter of the largest reinforcement in the topping slab. Previously, the only requirement was that the topping slabs must not be less than 2-1/2 in. thick.

The new provisions also specify that prestressing tendons shall not be used as primary reinforcement in boundaries and collector elements of structural diaphragms, and that precompression from unbonded tendons may be used to resist diaphragm forces.

New requirements on the reinforcement details for chords and collectors at splices and anchorage zones at the boundaries of structural diaphragms are contained in 1921.6.7.3.

1.3.6 Frame Members Not Part of the LFR System

In accordance with the revised deformation compatibility provisions (see Sect. 1.2.11), changes have been made in the design and detailing requirements for members that are not part of the LFR system (1921.7).

Section 1921.7.1 requires that frame members which are not part of the LFR system must be detailed according to 1921.7.2 or 1921.7.3, depending on the magnitude of the moments induced in those members when subjected to the maximum inelastic response displacement (Δ_M). The load combinations to be used in this case are $1.4D + 1.4L + E$ and $0.9D + E$. Previously, the moments were determined based on the members being subjected to $3R_w/8$ times the displacement resulting from the specified lateral forces.

In general, more restrictive detailing requirements are specified in 1921.7. This is mainly in response to the collapse or partial collapse of at least two parking garages during the 1994 Northridge earthquake where interior columns designed to carry gravity loads only failed.

1.4 REFERENCES

1.1 *Uniform Building Code*, International Conference of Building Officials (ICBO), Whittier, CA, 1997.

1.2 *Building Code Requirements for Structural Concrete (ACI 318-95) and Commentary (ACI 318R-95)*, American Concrete Institute, Farmington Hills, MI, 1995.

1.3 *Uniform Building Code*, International Conference of Building Officials (ICBO), Whittier, CA, 1994.

2 SEISMIC DESIGN LOADS

2.1 INTRODUCTION

This chapter outlines the general requirements for the analysis and design of buildings subjected to earthquake forces according to the provisions of the 1997 UBC. The applicable load combinations of the strength design method are given in 1612. Sections 1626 through 1636 contain the earthquake design provisions, including methods of determining the minimum design lateral forces due to the design basis ground motion (i.e., ground motion that has a 10% chance of being exceeded in 50 years). In general, these forces are determined considering the building's location relative to the seismic source zone or fault, site characteristics, occupancy, configuration, structural system, and height.

2.2 LOAD COMBINATIONS

As was noted in Chapter 1, one of the major changes in the 1997 UBC is the use of a strength-level design base shear, the effects of which are not to be amplified by load factors before combining with factored gravity load effects. Accordingly, the load combinations of 1612.2.1 are essentially those contained in Sect. 2.3 of ASCE 7-95 (Ref. 2.1), with some minor modifications.

It is important to note the exceptions at the end of 1612.2.1 that pertain to concrete and masonry. In particular, the load combinations of 1909.2 are applicable for concrete where the load combinations do not include seismic forces. Also, for concrete and masonry, the load combinations of 1612.2.1 that include seismic forces must be multiplied by 1.1. Table 2-1 contains the applicable load combinations for concrete subjected to the combined effects of dead (D), live (L), snow (S), wind (W), and earthquake forces (E).

Table 2-1 Load Combinations for Concrete Members

Equation No.	Load Combination*
(9-1)	$1.4D + 1.7L$
(9-2)	$0.75(1.4D + 1.7L + 1.7W)$
(9-3)	$0.9D + 1.3W$
(12-5)	$1.1(1.2D + f_1L + f_2S + 1.0E)$
(12-6)	$1.1(0.9D + 1.0E)$

*D = *effect of dead loads*
L = *effect of live loads*
W = *effect of wind loads*
E = *effect of earthquake loads*
f_1 = *1.0 for floors in places of public assembly, for live loads in excess of 100 psf, and for garage live loads*
= *0.5 for other live loads*
f_2 = *0.7 for roof configurations that do not shed snow off the structure*
= *0.2 for other roof configurations*

In addition to the basic load combinations, the special seismic load combinations of 1612.4 must also be considered:

$$1.2D + f_1 L + 1.0E_m$$

$$0.9D + 1.0E_m$$

where E_m is the estimated maximum earthquake force (see Sect. 2.4.1). These combinations are used to establish design loads for elements of a structure that are expected or required to remain elastic during the design ground motion. For example, in the case of columns supporting discontinued shearwalls, the columns must have the strength to resist the maximum load that can develop in the system.

The Structural Engineers Association of California (SEAOC) Seismology Committee has recently submitted specific recommendations to the International Conference of Building Officials (ICBO) regarding the second exception contained in 1612.2.1, based on the information in Ref. 2.2. In short, the Seismology Committee recommends that the 1.1 multiplier not be applied to Eq. (12-5) or (12-6). The reader is urged to check with the appropriate local jurisdiction to determine if this recommendation has been adopted or not. Reference 2.2 contains the background information that led to this recommendation.

2.3 Criteria Selection

2.3.1 Occupancy Categories

Each building is to be placed in one of the five occupancy categories in Table 16-K. This table contains the seismic importance factor (I) that is used in computing the seismic forces on the main lateral force resisting (LFR) system, and the seismic importance factor (I_p) that is used in determining the lateral forces on the elements of structures, nonstructural components, and equipment supported by structures.

2.3.2 Site Geology and Soil Characteristics

Local soil conditions can play a key role in the amplification of ground motion, and, thus, can dramatically effect the overall performance of a building during an earthquake. Essentially, the seismic design forces are magnified for different soil conditions. Based on recent studies, a new set of six soil profile categories has been introduced which has replaced the previous set of four categories. Soil profile types S_A, S_B, S_C, S_D, and S_E are defined in Table 16-J; soil profile type S_F is defined as soils requiring site-specific evaluation (see 1629.3.1). A procedure for determining the soil profile type for a given site is contained in 1636. Note that type S_D is to be used when the soil properties are not known in sufficient detail to determine the soil profile type.

2.3.3 Site Seismic Hazard Characteristics

Seismic Zone. The seismic zone map of the United States is shown in Fig. 16-2. Note that this map is based on a 10 percent probability of the ground motion being exceeded in a 50-year period. The seismic zone factors (Z) contained in Table 16-I represent the effective peak ground accelerations corresponding to the maximum expected level of earthquake ground motion for a given zone, and are expressed as a fraction of the gravity constant (g). Appendix Chapter 16 contains seismic zones for areas outside of the United States.

Near-Source Factors. The recognition of near-source effects in Seismic Zone 4 is a major change that was introduced into the 1997 UBC. Observations and readings from recent earthquakes indicate that ground motion effects close to a fault can be significantly greater than those recorded away from the fault rupture. The near-source factors (N_a) and (N_v), which were introduced to account for this phenomenon, are based on the seismic source type (which is related to the maximum moment magnitude and slip rate; see Table 16-U) and the distance of the site to the known seismic source. Figure 2-1 depicts the distance to the seismic source as defined in Footnote 3 of Tables 16-S and 16-T.

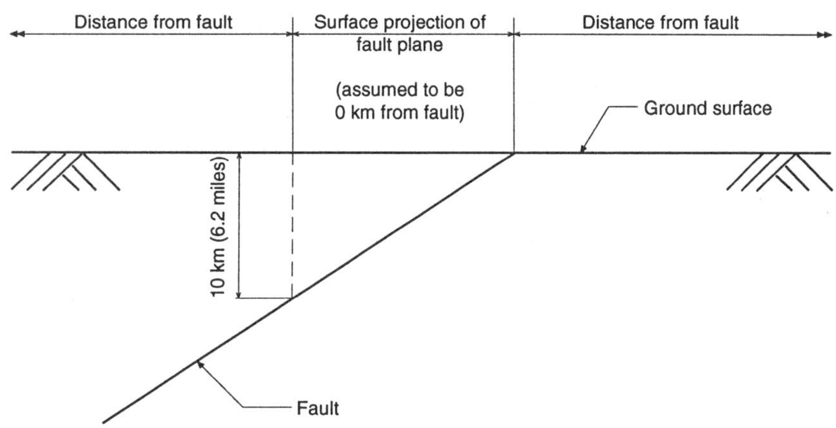

Figure 2-1 Distance from Known Seismic Source to be used in Determining N_a and N_v

Tables 16-S and 16-T contain the values of N_a and N_v, respectively, as a function of the source type and distance to the seismic source. Maps prepared by the United States Geological Survey or the California Division of Mines and Geology can be used for locating known faults.

Seismic Response Coefficients. The seismic coefficients C_a and C_v in Tables 16-Q and 16-R, respectively, are specified in terms of the seismic zone factor (Z), the soil profile type, and for Seismic Zone 4, the near-source factors (N_a) and (N_v). Note that for soil profile type S_F, a site-specific geotechnical investigation and dynamic site response analysis must be performed to determine C_a and C_v.

2.3.4 Building Configuration

The configuration of a building can significantly affect its overall performance during an earthquake. According to 1629.5.1, buildings shall be classified as regular or irregular based on plan and vertical configurations. Five different types of vertical structural irregularities are defined in Table 16-L (see Fig. 2-2): stiffness irregularity – soft story, weight (mass) irregularity, vertical geometric irregularity, in-plane discontinuity in vertical lateral force-resisting elements, and discontinuity in capacity – weak story. Five different types of plan structural irregularities are defined in Table 16-M (see Fig. 2-3): torsional irregularity, re-entrant corners, diaphragm discontinuity, out-of-plane offsets, and nonparallel systems.

Past earthquakes have shown that buildings having irregular configurations are likely to suffer greater damage than buildings having regular configurations.

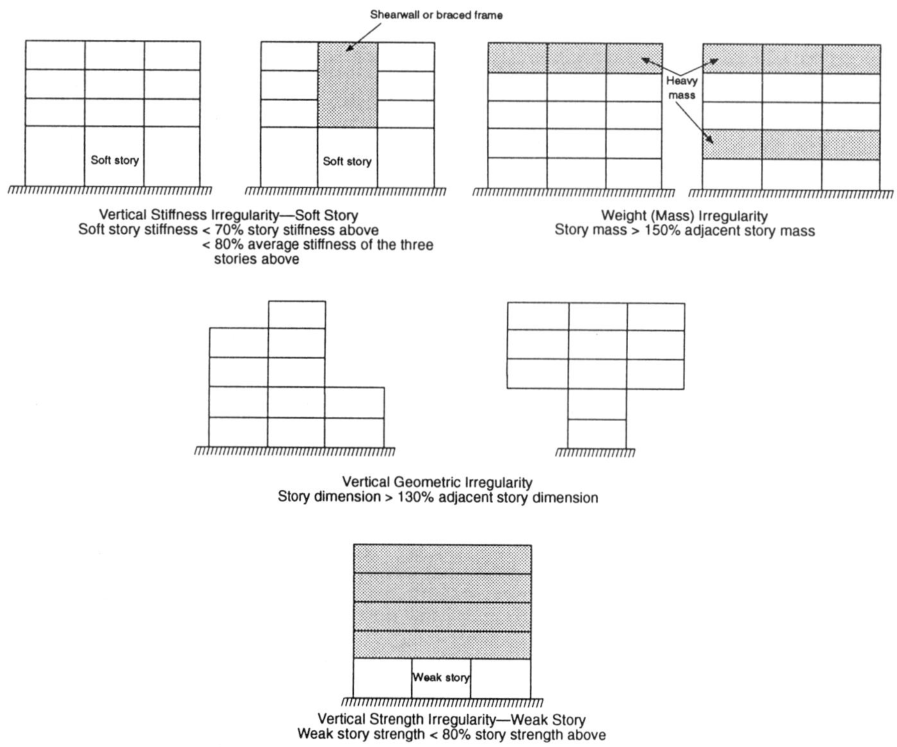

Figure 2-2 Vertical Structural Irregularities

2.3.5 Structural Framing Systems

The basic structural systems contained in Table 16-N are depicted in Fig. 2-4. It is important to note the system and building height limitations given in the table. A brief description of each system follows.

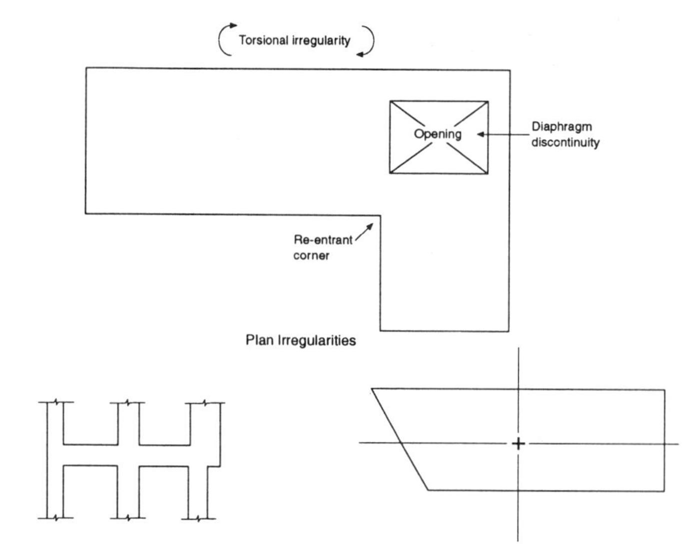

Figure 2-3 Plan Structural Irregularities

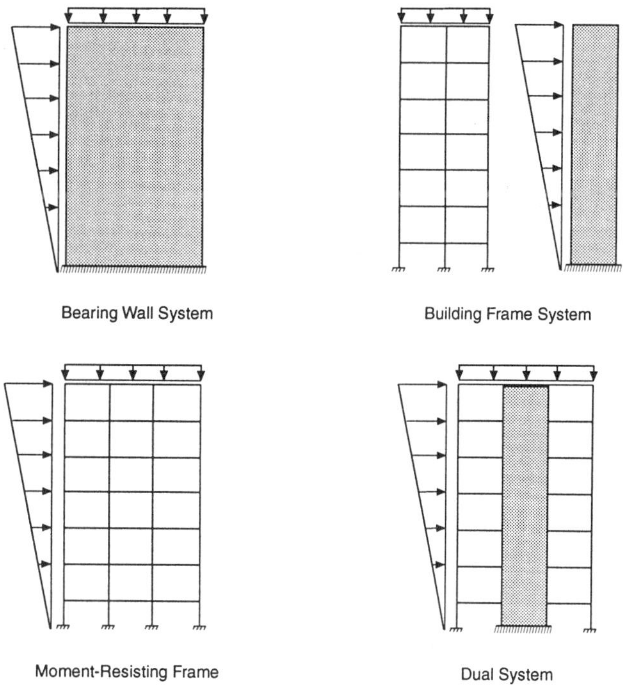

Figure 2-4 Earthquake-Resisting Structural Systems of Reinforced Concrete

Bearing Wall System. Loadbearing walls provide support for most or all of the gravity loads. Resistance to lateral forces are provided by the same walls acting as shearwalls. In Seismic Zones 3 and 4, the walls must be specially detailed according to the provisions of 1921.6.6 in addition to the applicable requirements in 1901 through 1917. In lower seismic zones, no special detailing of the walls is required.

Building Frame System. An essentially complete space frame provides support for the gravity loads. Resistance to lateral forces is provided by shearwalls. In Seismic Zones 3 and 4, the shearwalls must satisfy the provisions of 1921.6.6 in addition to the applicable requirements in 1901 through 1917. No special detailing of the shearwalls is required in lower seismic zones. Although the walls are designed to carry all of the seismic forces, the beam-column frames must be designed to resist the effects caused by the lateral load deflections, since they are connected to the walls at every floor level through the floor slabs. This is the so-called deformation compatibility requirement for buildings in Seismic Zones 2, 3, and 4 (1633.2.4). According to this provision, members not designated to be part of the lateral force resisting (LFR) system must be capable of maintaining support of the gravity loads when subjected to the expected deformations caused by the seismic forces. Special requirements for concrete members are contained in 1921.7. If it is determined by the provisions of 1921.7 that special seismic details are required in structural members originally designed for gravity only, then the designer should consider making these members part of the LFR system.

In regions of low seismic activity (Seismic Zones 0 and 1), it is not practical or economical to design the shearwalls for 100% of the lateral loads, since no special seismic detailing is required for the frame members. Thus, in Seismic Zones 0 and 1, structures containing shearwalls and frames should be designed as shearwall-frame interaction systems (see discussion below).

Moment-resisting Frame System. An essentially complete space frame provides support for the gravity loads. Resistance to lateral loads is provided by moment-resisting frames. In Seismic Zones 0 and 1, the moment-resisting frames can be ordinary moment-resisting frames (OMRF) which do not require any special seismic detailing; i.e., the structural members must satisfy the provisions of 1901 through 1918. Intermediate moment-resisting frames (IMRF) proportioned to satisfy the requirements of 1921.8 in addition to those in 1901 through 1918 must be provided in Seismic Zone 2 (1633.2.7). In Seismic Zones 3 and 4, special moment-resisting frames (SMRF) are required; these frames must be designed and detailed according to 1921.2 through 1921.7, in addition to 1901 through 1917.

Dual System. This type of structural system has three fundamental features: (1) an essentially complete space frame provides support for the gravity loads, (2) resistance to lateral loads is provided by shearwalls and moment-resisting frames capable of independently resisting at least 25% of the base shear, and (3) the shearwalls and moment-resisting frames are designed to resist the design base shear in proportion to their relative rigidities. In Seismic Zones 3 and 4, the moment frames must be SMRF as defined earlier and the shearwalls must conform to the provisions of 1921.6.6 in addition to the applicable requirements in 1901 through 1917. In Seismic Zone 2, the moment frames must be IMRF, and no special details are required for the shearwalls. The moment frames can be OMRF in Seismic Zones 0 and 1; also, the shearwalls do not require any special detailing.

In addition to the four basic systems, two new structural system classifications have been introduced in the 1997 UBC: (1) cantilevered column building systems and (2) shearwall-frame interaction systems. A cantilevered column building system consists of a structure supported on columns that are fixed at the base and that have minimal moment capacity at the top. In addition to supporting the gravity loads, these column elements provide resistance to the lateral loads that are essentially applied at the top of the column. This type of structure has limited redundancy, since failure of the columns would result in failure of the gravity load carrying capacity as well. Typically, this system can be used to simulate a weak story in a building.

A shearwall-frame interaction system uses combinations of shearwalls and frames to resist the lateral loads. Interaction between the walls and frames is considered at every level of the structure, and the lateral forces are allocated to these elements based on their relative rigidities. It is important to note that this type of structural system is to be used in regions of low seismic activity, i.e., Seismic Zones 0 and 1 (see footnote 8 of Table 16-N). Thus, no special detailing is required for the shearwalls or the ordinary moment-resisting frame. As was noted above, it is not practical or economical to design the shearwalls to resist all of the lateral load effects in regions of low seismic activity, since no special detailing is required for the frame members.

The response modification factor (R) contained in Table 16-N represents the ratio of forces that would develop in a structure under the design earthquake if it behaved in an entirely elastic manner to the code-prescribed design forces. Since R is always greater than 1, all structures are designed for forces less than those that the design earthquake would produce in a completely linear-elastic structure (it is generally uneconomical and also impractical to design a structure to respond in the elastic range to the maximum probable design forces). This reduction is based on the premise that a properly detailed structure will respond inelastically and will be able to dissipate the excess earthquake energy without collapse. The R-factor is larger for structures that have a larger toughness (i.e., a greater capability to dissipate energy).

2.4 Minimum Design Lateral Forces and Related Effects

Both static and dynamic procedures are provided in the 1997 UBC to determine the minimum seismic forces acting on the LFR system of a building. Any structure may be designed according to the dynamic procedures of 1631; however, it is important to note that structures that are not included in 1629.8.3 and those that are included in 1629.8.4 must be designed using one of the dynamic procedures.

New to the 1997 UBC is a simplified static method to determine the seismic forces (1629.8.2). This method is applicable to (1) buildings that use light-frame construction of any occupancy (including single-family dwellings) not more than 3 stories in height, excluding basements, and (2) other buildings not more than 2 stories in height, excluding basements. Code sections that do not apply when using the simplified method are contained in 1630.2.3.4.

The following discussion will focus on the determination of the seismic forces using the static force procedure contained in 1630.

2.4.1 Earthquake Loads

According to 1630.1.1, earthquake loads (E) and (E_m) are to be used in the load combinations specified in 1612 (see Sect. 2.2). The earthquake load (E) is a function of both the horizontal and vertical components of the ground motion, and is determined from Eq. (30-1):

$$E = \rho E_h + E_v$$

where E_h = earthquake force due to the base shear (V)

E_v = the load effect resulting from the vertical component of the earthquake motion

$= 0.5C_aID$

ρ = reliability/redundancy factor (see Eq. 30-3)):

$$1.0 \le \rho = 2 - \frac{20}{r_{max}\sqrt{A_B}} \le 1.25 \text{ for special moment-resisting frames}$$

$$\le 1.5 \text{ for other systems}$$

As its name implies, the new factor (ρ) is essentially a measure of the redundancy of the structure. In general, it is a function of the number of lateral force resisting elements in the structure, the lateral load distribution in these elements, and the plan area of the building. The quantities (r_{max}) and (A_B) are the maximum element-story shear ratio and the ground floor area of the structure in square feet, respectively. To account for the inherent redundancy in dual systems, the redundancy factor may be taken as 80 percent of the value calculated by Eq. (30-3). Also, for structures located in Seismic Zones 0, 1, or 2, or, when calculating drift, ρ shall be taken equal to 1.0.

According to 1630.1.1, the maximum element-story shear ratio (r_{max}) is defined as the largest of the element-story shear ratios (r_i) which occurs in any of the story levels at or below the two-thirds height level of the building. For a given direction of loading, r_i is the ratio of the design story shear in the most heavily loaded single element divided by the total design story shear. Table 2-2 contains the definitions of r_i for various framing systems.

Table 2-2 Definition of Element-Story Shear Ratios

Structural system	Element-story shear ratio (r_i)
Moment frame	Maximum of the sum of the shears in any two adjacent columns in a bay divided by the story shear. For columns common to two bays in the direction under consideration, 70% of the shear in that column may be used in the column shear summation. The bay with the largest total shear force is used to determine r_i.
Shearwalls	Maximum value of the product of the wall shear multiplied by $10/\ell_w$ and divided by the total story shear, where ℓ_w is the length of the wall in feet. The wall with the largest shear force is used to determine r_i.
Dual systems	Ratio of the design story shear in the most heavily loaded single element (considering all lateral load resisting elements) divided by the total design story shear.

The magnitude of ρ can obviously have a dramatic influence on the effects of the horizontal earthquake design forces; an increase of the design forces on the structural members by up to 50% can be realized for a structural system that lacks a sufficient amount of redundancy.

The Structural Engineers Association of California (SEAOC) Seismology Committee has recently submitted specific recommendations to the International Conference of Building Officials (ICBO) regarding the use of E_v in Eqs. (12-5) and (12-6), based on the information in Ref. 2.2. In short, the Seismology Committee recommends that E in Eq. (12-5) be taken as that which is defined by Eq. (30-1), but that E in Eq. (12-6) be taken equal to E_h. The reader is urged to check with the appropriate local jurisdiction to determine if this recommendation has been adopted or not. Reference 2.2 contains the background information that led to this recommendation.

The parameter E_m is the estimated maximum earthquake force that can be developed in the structure (Eq. 30-2):

$$E_m = \Omega_o E_h$$

where Ω_o is the seismic force amplification factor for a structural system contained in Table 16-N. Note that Ω_o is a representation of the structural overstrength, comparable to the $3R_w/8$ factor which was in the 1994 UBC. As was discussed in Sect. 2.2, the overstrength factor is used in the special seismic load combinations to establish design loads for specific elements of the structure that are expected or required to remain elastic during the design ground motion (e.g., columns supporting a discontinued shearwall).

2.4.2 Modeling Requirements

Guidelines for the creation of the structural model for lateral load analysis are given in 1630.1.2. For concrete and masonry structures, cracked sections must be used in the model. For structures where deformation compatibility must be considered, the flexural and shear stiffness properties shall not exceed one-half of the gross section properties for concrete and masonry elements that are part of the lateral force resisting system, unless a rational cracked section analysis is performed (1633.2.4).

2.4.3 P-Δ Effects

Seismic forces cause a structure to deflect laterally. As a result, secondary moments are induced in the structural members due to the displaced gravity loads as shown in Fig. 2-5. This secondary effect is known as the P-Δ effect. According to 1630.1.3, P-Δ effects shall be evaluated using the design seismic forces producing the design level response displacements (Δ_S) (see 1630.9.1 and Sect. 2.4.7). In general, P-Δ effects need not be considered when the stability coefficient (θ), defined as the ratio of secondary moments to primary moments, is less than or equal to 0.10. The stability coefficient (θ) for a given story can be computed from the following equation:

$$\theta = \frac{P_x \Delta}{V_x h_{sx}}$$

where P_x = total unfactored gravity load at and above level x

Δ = seismic story drift

V_x = seismic shear force between levels x and $x - 1$

h_{sx} = story height below level x

In Seismic Zones 3 and 4, P-Δ effects need not be considered when the story drift (Δ) is less than or equal to $0.02h_{sx}/R$.

Flexible structures and structures with larger R values are more likely to be effected by P-Δ effects. Note that concrete columns need not have their bending moments amplified according to the provisions of 1910.11 when a P-Δ analysis is performed, since it is the intent of 1910.11 to account for P-Δ effects.

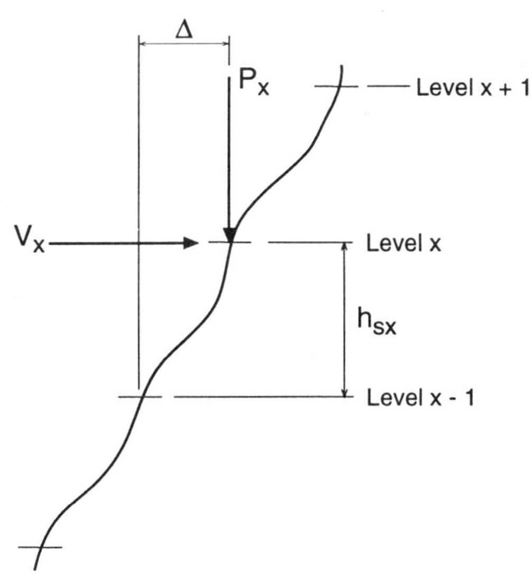

Figure 2-5 P-Δ Effects

2.4.4 Static Force Procedure

Seismic Base Shear. The static force procedure is an approximate method of determining the seismic base shear. For structures that conform to the provisions of 1629.8.3, the total design seismic base shear (V) in a given direction can be computed from Eq. (30-4), considering the limitations in Eqs. (30-5) through (30-7):

$$\frac{2.5C_a IW}{R} \geq V = \frac{C_v IW}{RT} \geq 0.11C_a IW$$

$$\geq \frac{0.8ZN_v IW}{R} \quad \text{(Seismic Zone 4)}$$

where C_a and C_v = seismic coefficients contained in Tables 16-Q and 16-R, respectively

I = seismic importance factor (Table 16-K)

W = total dead load and applicable portions of other loads as specified in 1630.1.1

R = response modification factor (Table 16-N)

T = fundamental period of vibration of the structure

Z = Seismic Zone Factor (Table 16-I)

N_v = velocity-dependent near-source factor (Table 16-T)

The minimum design base shear limitation for Seismic Zone 4 was introduced as a result of the ground motion effects observed at sites near the fault rupture in the 1994 Northridge earthquake.

Seismic Dead Load (W). In general, W is the total weight of the building and that part of the service load that will most likely be present at the time of an earthquake. Applicable parts of other loads that must be included in W are the following:

1. In storage and warehouse occupancies, a minimum of 25 percent of the floor live load.
2. Where partition loads are used in the floor design, a load of not less than 10 psf.
3. The snow load, when it is greater than 30 psf (with a possible reduction of up to 75 percent when approved by the building official).
4. Total weight of permanent equipment.

Approximate Fundamental Period (T). In order to compute the fundamental period of vibration of a building, member sizes must be known. Without member sizes, it becomes necessary to estimate the period in order to calculate the base shear, so that a preliminary design can be performed. Equation (30-8) of Method A in 1630.2.2 gives an estimate for the fundamental period (T):

$$T = C_t (h_n)^{3/4}$$

where C_t = 0.030 for moment-resisting frames of concrete

= 0.020 for other concrete building systems

h_n = height from the base of the building to the highest level, ft

Alternatively, for structures with concrete or masonry shearwalls:

$$C_t = \frac{0.1}{\sqrt{A_c}}$$

where A_c = combined effective area of shearwalls in the first story of the structure, ft^2

$$= \sum A_e[0.2 + (D_e / h_n)^2],\ D_e / h_n \leq 0.9$$

A_e = minimum cross-sectional area in any horizontal plane in the first story of a shearwall, ft^2

D_e = length of a shearwall in the first story in the direction parallel to the applied forces, ft

In lieu of Method A, Method B can be used to determine T. Method B permits the evaluation of T by either the Rayleigh formula (Eq. (30-10)) or any other substantiated analysis. Note that the value of T obtained from Method B must be less than or equal to 1.3 times the value of T obtained from Method A in Seismic Zone 4, and less than or equal to 1.4 times the value in Seismic Zones 1, 2, and 3.

Once preliminary member sizes are obtained based on the base shear computed using the approximate period, a more accurate value of T can be determined using established procedures.

Comparisons between the base shears from the 1994 and 1997 UBC provisions are given in Chapter 1.

2.4.5 Distribution of Lateral Forces

The base shear (V) is first distributed vertically to each floor level of the building. The story shears at each floor level are then distributed to the various LFR elements depending on their relative stiffnesses and the stiffness of the diaphragm. The procedures for vertical and horizontal seismic force distributions are explained below.

Vertical Force Distribution. According to 1630.5, the total base shear shall be distributed over the height of the building in conformance with Eqs. (30-13) through (30-15). Figure 2-6 summarizes the distribution of the lateral forces according to these equations. A portion of the base shear is concentrated at the top of buildings with fundamental periods greater than 0.7 seconds to account for higher mode effects. The remaining portion of the base shear is distributed linearly over the height of the building, varying from a maximum value at the top to a minimum value at the bottom, in correspondence with the fundamental mode response.

Horizontal Force Distribution. The seismic design story shear (V_x) in any story is determined as follows (1630.6):

$$V_x = F_t + \sum_{i=x}^{n} F_i$$

where F_i is the portion of the seismic base shear (V) located at or assigned to level i. The seismic design story shears are distributed to the various LFR elements in proportion to their rigidities and the rigidity of the diaphragm. According to 1630.6, a diaphragm is considered to be flexible when the maximum lateral deformation of the diaphragm is more than two times the average story drift of the associated story. In the case of a concrete slab, it is unlikely that it will be anything but a rigid diaphragm unless it has a highly irregular shape and/or has large openings.

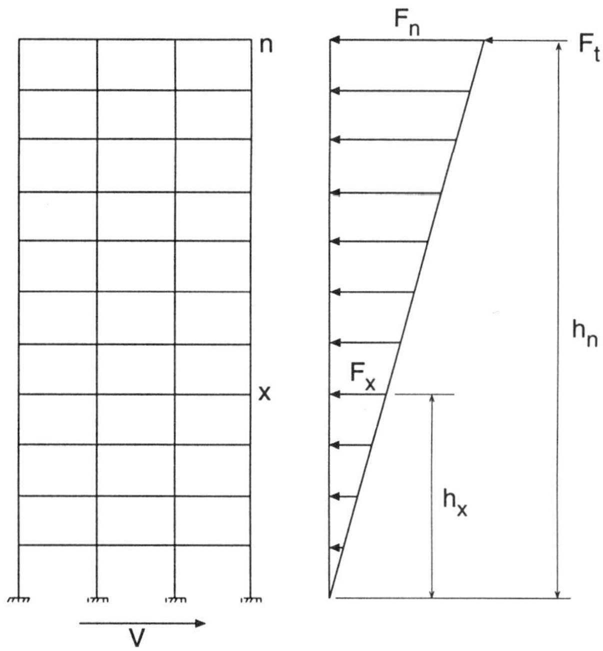

Vertical Force Distribution

$$V = F_t + \sum_{i=1}^{n} F_i$$

$$F_t = 0.07TV \le 0.25V \quad \text{for } T > 0.7 \text{ seconds}$$
$$= 0 \quad \text{for } T \le 0.7 \text{ seconds}$$

$$F_x = \frac{(V - F_t) w_x h_x}{\sum_{i=1}^{n} w_i h_i}$$

F_i, F_n, F_x = *design seismic force applied to level i, n, or x, respectively*

F_t = *portion of V considered concentrated at top of structure in addition to* F_n

h_i, h_n, h_x = *height in ft above base to level i, n, or x, respectively*

w_i, w_x = *that portion of W which is located at or assigned to level i or x, respectively*

W = *total seismic dead load*

T = *fundamental period of vibration of structure in seconds in direction of analysis*

V = *total design base shear*

Figure 2-6 Distribution of Design Lateral Forces Along Building Height

When the diaphragm is determined to be not flexible, provisions shall be made to account for the additional forces induced in some of the LFR elements due to torsion of the building. Section 1630.7 specifies that the torsional design moment at a given story shall be the torsional moments resulting from any eccentricity between the center of mass and center of rigidity (resistance) plus an accidental torsional moment. It is assumed that the accidental torsional moment is caused by displacement of the center of mass each way from its actual location by a distance equal to 5 percent of the dimension of the building perpendicular to the direction of the applied forces.

When torsional irregularity as defined in Table 16-M exists, the torsional effects shall be accounted for by increasing the accidental torsion at each level by a torsion amplification factor (A_x) determined from Eq. (30-16):

$$A_x = \left(\frac{\delta_{max}}{1.2\delta_{avg}}\right)^2 \le 3.0$$

where δ_{max} = maximum displacement at level x

δ_{avg} = average of the displacements at the extreme points of the structure at level x

2.4.6 Overturning

Buildings must be designed to resist the overturning effects caused by the earthquake forces (1630.8). The overturning moment (M_x) at any level x can be determined from the following formula:

$$M_x = \sum_{i=x}^{n} F_i(h_i - h_x) + F_t(h_n - h_x)$$

where F_i = portion of V located at or assigned to level i

F_t = portion of V located at the top of the structure in addition to F_n

h_i, h_n, h_x = height from the base to level i, n, or x

The overturning moments are distributed to the elements of the LFR system in the same manner as the horizontal shear forces (see 1630.6).

2.4.7 Story Drift Limitations

According to 1630.10, story drifts shall be computed using the maximum inelastic response displacement (Δ_M), which is an approximation of the displacement that occurs when the structure is subjected to the design basis ground motion:

$$\Delta_M = 0.7R\Delta_S$$

The design level response displacements (Δ_S) due to the code-prescribed seismic forces from 1630.2.1 (see Sect. 2.4.4) are determined at each story level based on a static, elastic analysis of the LFR system, including torsional effects (1630.9.1). Alternatively, the displacements may be computed using an elastic dynamic analysis in accordance with 1631. In any case, the modeling requirements of 1630.1.2 must be satisfied.

In lieu of the above equation, Δ_M may be computed by a nonlinear time history analysis, which will typically provide a more reliable estimation of the inelastic drift (see 1631.6).

If it is determined by the provisions of 1630.1.3 that a P-Δ analysis is required, Δ_M must include displacements associated with the second-order effects.

For structures with a fundamental period (T) less than 0.7 seconds, the calculated story drift using Δ_M shall not exceed 0.025 times the story height. For structures with T greater than or equal to 0.7 seconds, the story drift shall not exceed 0.020 times the story height. It is important to note that these drift limits are appropriate for structures analyzed according to the static force procedure and response spectrum techniques. Appropriate drift limits are recommended in Appendix C Commentary of Ref. 2.3 when more sophisticated analysis techniques are utilized.

For the purposes of drift analysis only, the design lateral forces used to determine the calculated drift may disregard the limitation of Eq. (30-6). Also, when using Eq. (30-10) to determine the period (T), the 30 or 40 percent limitations prescribed in 1630.2.2 may be neglected.

2.5 Detailed Systems Design Requirements

In general, all structural framing systems that resist seismic forces must comply with the requirements of 1629. For concrete buildings, the components of these systems must be designed for the effects of the seismic forces according to the provisions of Chapter 19. Additionally, framing systems and components must comply with 1633. The following discussion focuses on requirements that typically must be taken into account in the design of concrete structures.

2.5.1 Orthogonal Effects

In Seismic Zones 2, 3, and 4, the effects of earthquake forces acting in a direction other than the principle axes must be considered in each of the following cases (1633.1):

(1) the structure has plan irregularity Type 5 (Table 16-M)

(2) the structure has plan irregularity Type 1 (Table 16-M) for both major axes

(3) a column of a structure is part of two or more intersecting LFR systems

Note that if the axial load in a column due to seismic forces acting in either direction is less than 20 percent of the column axial load capacity, orthogonal effects need not be considered.

Two methods may be used to satisfy the requirement of orthogonal effects: (1) design the element for 100 percent of the design seismic forces in one direction plus 30 percent of the design seismic forces in the perpendicular direction, or (2) combine the effects of the two orthogonal directions using the square root of the sum of the squares (SRSS). When using the SRSS method, each term computed shall be assigned the sign that will result in the most conservative result.

2.5.2 Deformation Compatibility

For buildings in Seismic Zones 2, 3, and 4, elements of a structure not designated to be part of the LFR system must be designed and/or detailed to maintain support of the gravity loads when subjected to the expected deformations caused by the seismic forces (1633.2.4). Note that P-Δ effects on these elements shall be considered. The expected deformation is the greater of the maximum inelastic response displacement (Δ_M) determined by 1630.9.2 or the deformation induced by a story drift of 0.0025 times the story height. The forces induced in these elements by the expected deformation may be considered ultimate or factored forces.

The assumed flexural and shear stiffness properties for concrete and masonry elements that are part of the LFR system shall not exceed one-half of the gross section properties unless a rational cracked-section analysis is performed. Special requirements for concrete members that are not part of the LFR system are contained in 1921.7.

A building frame system is one type of structural system where deformation compatibility is to be considered. The beam-column frames must be designed to resist the effects of the expected deformations, since they are connected to the walls (which are designed to carry all of the seismic forces) at every floor level through the floor slabs. The designer should consider changing from a building frame system to another structural system (e.g., a dual system) if it is determined by the provisions of 1921.7 that special seismic details are required in members originally designed for gravity loads only. By doing so, a more efficient design may be realized.

2.5.3 Diaphragms

Floor and roof diaphragms shall be designed to resist the forces (F_{px}) determined from Eq. (33-1):

$$1.0C_a I w_{px} \geq F_{px} = \frac{F_t + \sum_{i=x}^{n} F_i}{\sum_{i=x}^{n} w_i} w_{px} \geq 0.5 C_a I w_{px}$$

where w_{px} is the weight of the diaphragm and the tributary elements thereto at level x, including the applicable portions of other loads as defined in 1630.1.1, and all other terms have been defined previously. Design and detailing requirements for concrete diaphragms are contained in 1633.2.9 and 1921.6.

2.6 References

2.1 ASCE Standard *Minimum Design Loads for Buildings and Other Structures* (ASCE 7-95), American Society of Civil Engineers, New York, 1996.

2.2 Ghosh, S.K., "Design of Reinforced Concrete Buildings under the 1997 UBC," *Building Standards*, May-June, International Conference of Building Officials, Whittier, CA, 1998, pp. 20-24.

2.3 Seismology Committee of the Structural Engineers Association of California, *Recommended Lateral Force Requirements and Commentary*, Structural Engineers Association of California, Sacramento, CA, 1996.

3 Detailing Requirements

3.1 Introduction

The seismic design forces determined by the methods in Chapter 2 are only a fraction of the actual forces that an elastic structure may experience during an earthquake. For design purposes, the seismic forces are reduced since it is impractical as well as uneconomical to design a structure to remain elastic. However, in order to prevent collapse, structures must be properly detailed so that they are able to dissipate the excess earthquake energy through inelastic deformations.

In general, the detailing requirements of the structural members in a building depend on the severity of the seismic zone. The 1997 UBC provisions of 1921 for the seismic design of reinforced concrete members are taken from Chapter 21 of ACI 318-95 (Ref. 3.1), with some modifications.

Note that Chapter 21 of ACI-318 specifies "low," "moderate," and "high" seismic risk levels. These seismic risk levels are based on equal probability of risk of damage and are related to the intensity of the ground shaking in the same manner as the seismic zones.

Table 3-1 relates the seismic risk levels referred to in ACI 21.2.1 and the seismic zones defined in 1921.2.1. The proportioning and detailing requirements are contained in Table 3-2. Table 3-3 summarizes the proportioning and detailing requirements for moment frames.

Table 3-1 Seismic Risk Terminologies in ACI 318-95 and UBC-97

ACI	Low	Moderate	High
UBC	Zones 0 & 1	Zones 2A & 2B	Zones 3 & 4

It is important to note that this chapter does not attempt to present each and every requirement specified in the UBC. For complete or additional information, the appropriate sections in the UBC should be consulted (Ref. 3.2).

Table 3-2 Proportioning and Detailing Requirements

Seismic Risk Level	Zone	Proportion and Detail by UBC		
		1901 - 1918	1901 – 1918 plus 1921.8	1901 - 1917 plus 1921 (excluding 1921.7, 1921.8)
Low	0 & 1	Beam-Column or Slab-Column Frames, Shearwalls		
Moderate	2A & 2B	Shearwalls	Beam-Column or Slab-Column Frames	
High	3 & 4			Beam-Column Frames, Shearwalls

Note: Requirements listed apply only to frames and shearwalls that are part of the lateral force resisting system. In moderate and high seismic zones, 1921.7 must be satisfied for frame members not proportioned to resist forces induced by earthquake motions.

Table 3-3 Proportioning and Detailing Requirements for Moment Frames

Frame Type	Proportion and Detail by
Ordinary Moment Frame	UBC, exclusive of 1921
Intermediate Moment Frame	UBC, exclusive of 1921 but inclusive of 1921.8
Special Moment Frame	UBC, inclusive of 1921.2 through 1921.5

3.2 General Requirements

3.2.1 Design and Detailing

Seismic Zones 0 and 1 (Regions of Low Seismic Risk). For concrete structures located in regions of low seismic risk (minor or no risk of damage), no special design or detailing is required; thus, the general requirements of 1901 through 1918, excluding 1921, apply. Concrete structures proportioned by the general requirements of the code are considered to have a level of toughness adequate for low intensity earthquakes.

It is important to note that the general requirements of the code include several provisions specifically intended to improve toughness, in order to increase the resistance of concrete structures to earthquake and other catastrophic or abnormal loads. For example, when a beam is part of the primary lateral force resisting (LFR) system of a structure, a portion of the positive moment reinforcement must be anchored at the supports to develop its yield strength in tension (1912.11.2). Also, hoop reinforcement must be provided in certain types of beam-column connections (1911.11.2). Other design provisions include minimum shear reinforcement (1911.5.5) and bar anchorage and splicing details (1912). The provisions of 1907 addressing special reinforcement for structural integrity enhance the overall integrity of concrete structures in the event of damage to a major supporting element or an abnormal load event.

Seismic Zones 2A and 2B (Regions of Moderate Seismic Risk). For concrete structures located in regions of moderate seismic risk (moderate risk of damage), 1921.8 includes requirements, in addition to those contained in 1901 through 1918, that are applicable to reinforced concrete moment frames (beam-column or slab-column framing systems). These requirements will provide a suitable level of inelastic behavior when the frame is subjected to an earthquake of such magnitude as to require it to perform inelastically. There are no design or detailing requirements in addition to those of 1901 through 1918 and 1921.8 for other structural components of buildings in regions of moderate seismic risk, including structural walls (shearwalls). Structural walls proportioned by the general requirements of the code are considered to have sufficient toughness at drift levels anticipated in regions of moderate seismicity.

The type of framing system provided for earthquake resistance in regions of moderate seismicity will dictate whether special reinforcement details need to be incorporated in the structure.

If the lateral force resisting (LFR) system consists of moment frames, intermediate details must be provided, even if the combinations including seismic loads do not govern. This is to ensure a minimum level of toughness in the moment resisting frames, since they are the only elements that resist the effects of an earthquake.

For frame-shearwall systems, inclusion of the intermediate details will depend on how the earthquake forces are assigned to the shearwalls and the frames. If the total earthquake forces are assigned to the shearwalls, frame members shall conform to 1921.7 (see 1921.2.1.3). If frame-shearwall interaction is considered in the analysis, then intermediate details are required for the frame portion of the dual framing system.

Seismic Zones 3 and 4 (Regions of High Seismic Risk). For concrete structures located in regions of high seismic risk (major risk of damage), all structural components must satisfy the applicable special proportioning and detailing requirements of 1921 (excluding 1921.8). If, for purposes of design, some of the frame members are not considered as part of the LFR system, special proportioning and detailing of these frame members by 1921.7 is still required (1921.2.2.4). The special provisions for seismic design of 1921 are intended to provide a monolithic reinforced concrete structure with adequate toughness to respond inelastically under severe earthquake motions. Note that precast concrete structural systems may be used if they satisfy the requirements of 1921.2.1.6 and 1921.2.1.7.

3.2.2 Strength-reduction Factors

According to 1921.2.3, the strength-reduction factors (ϕ) of 1909.3.4 are applicable for the design of the members.

3.2.3 Compressive Strength

The minimum concrete compressive strength shall not be less than 3,000 psi in Seismic Zones 3 and 4 (1921.2.4.1). This value is specified because it is felt that using a strength equal to this value and above will provide a concrete of acceptable quality. The compressive strength of lightweight aggregate concrete is limited to a maximum of 4,000 psi, unless it is demonstrated that structural members made with lightweight aggregate concrete have at least the same strength and toughness as members made with normal weight aggregate concrete of the same strength (1921.2.4.2). The limiting value for lightweight aggregate concrete is used because of limited data and research on this material for seismic applications.

3.2.4 Reinforcement

Reinforcement used in frame members and wall boundary elements shall comply with ASTM A706 (1921.2.5). ASTM A615 may be used provided the upper strength limits given in 1921.2.5.2 are satisfied. These upper limits are given to prevent a brittle failure caused by a higher than anticipated strength of the reinforcing steel.

3.3 MEMBERS SUBJECTED TO BENDING

Table 3-4 contains the requirements for flexural members. These requirements typically apply to beams of frames and other flexural members with negligible axial loads.

3.3.1 General

Flexural members must meet the general requirements of 1921.3.1.1 through 1921.3.1.4 so that the cross-sections are consistent with those that have performed well in the past. Members must have sufficient ductility and provide efficient moment transfer to the supporting columns. Note that columns subjected to bending and having a factored axial load $P_u \leq A_g f_c' / 10$ may be designed as flexural members, where A_g is the gross area of the section.

3.3.2 Flexural Reinforcement

The reinforcement requirements for flexural members in Seismic Zones 3 and 4 are shown in Fig. 3-1. The minimum moment capacity at any section of the beam is based on the moment capacity at the face of the support. These requirements ensure strength and ductility under large lateral displacements. The requirement that at least two bars be continuous at both the top and bottom of the section is for construction purposes.

The flexural requirements for members in Zone 2 are similar to those shown in Fig. 3-1 (see Table 3-4).

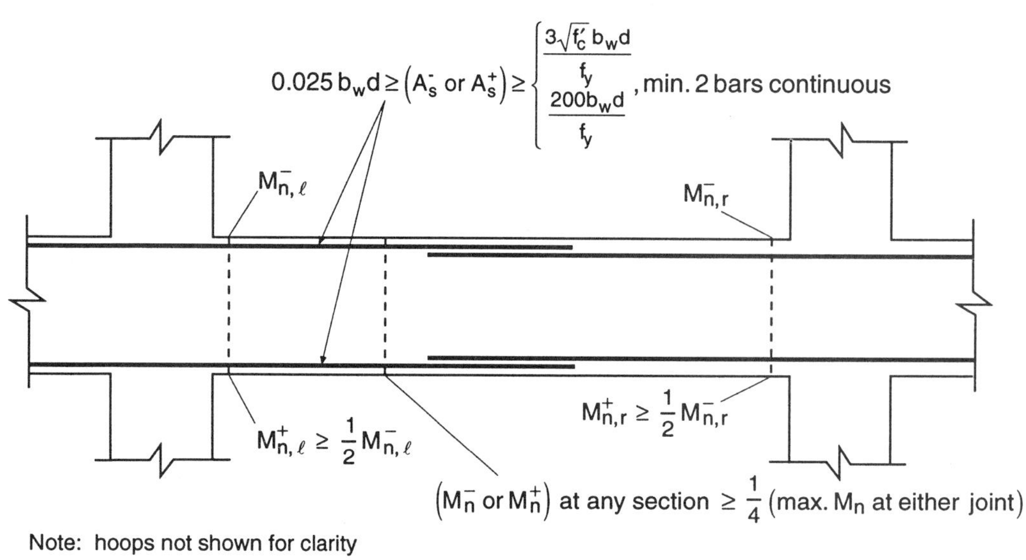

Figure 3-1 Reinforcement Requirements for Flexural Members in Seismic Zones 3 and 4

Table 3-4 Flexural Members of Frames

	Seismic Zones 3 & 4	Seismic Zones 0, 1 & 2
General	Flexural frame members shall satisfy the following conditions: • Factored axial compressive force ≤ $A_g f'_c$ / 10 • Clear span ≥ 4 × effective depth • Width to depth ratio ≥ 0.3 • Width ≥ 10 in. • Width ≤ width of supporting member + distances on each side of the supporting member not exceeding three-fourths of the depth of the flexural member **1921.3.1**	Zone 2 — Factored axial compressive force ≤ $A_g f'_c$ / 10. **1921.8.2** Zones 0 & 1 — No similar requirements.
Flexural Requirements	Minimum reinforcement shall not be less than $\dfrac{3\sqrt{f'_c}\,b_w d}{f_y}$ and $\dfrac{200\,b_w d}{f_y}$ at any section, top and bottom, unless provisions in 1910.5.3 are satisfied. **1921.3.2.1**	Same requirement except minimum reinforcement need only be provided at sections where tensile reinforcment is required by analysis, except as provided in 1910.5.2, 1910.5.3, and 1910.5.4. **1910.5**
Flexural Requirements	The reinforcement ratio (ρ) shall not exceed 0.025. **1921.3.2.1**	The reinforcement ratio (ρ) shall not exceed $0.75\rho_b$. **1910.3.3**
Flexural Requirements	At least two bars shall be provided continuously at both top and bottom of section. **1921.3.2.1**	Zones 0, 1 & 2 — Provide minimum structural integrity reinforcement. **1907.13**
Flexural Requirements	Positive moment strength at joint face ≥ 1/2 negative moment strength at that face of the joint. **1921.3.2.2**	Zone 2 — Positive moment strength at joint face ≥ 1/3 negative moment strength at that face of the joint. **1921.8.4.1** Zones 0 & 1 — No similar requirement.
Flexural Requirements	Neither the negative nor the positive moment strength at any section along the member shall be less than 1/4 the maximum moment strength provided at the face of either joint. **1921.3.2.2**	Zone 2 — Same requirement except only need to provide 1/5 of the maximum moment strength at the face of either joint at any section along the member. **1921.8.4.1** Zones 0 & 1 — No similar requirement.

A_g = gross area of section
b_w = width of web
d = effective depth of section
f'_c = compressive strength of concrete
f_y = yield strength of reinforcement
ρ_b = balanced reinforcement ratio

Table 3-4 Flexural Members of Frames (cont'd)

	Seismic Zones 3 & 4	Seismic Zones 0, 1 & 2
Splices	Lap splices of flexural reinforcement are permitted only if hoop or spiral reinforcement is provided over the lap length. Hoop and spiral reinforcment spacing shall not exceed *d*/4 or 4 in. Welded splices and mechanical connections shall conform to 1921.2.6. **1921.3.2.3, 1921.3.2.4**	There is no requirement that splices be enclosed in hoops. However, if closely spaced No. 4 hoops are provided, the lap splice length can be reduced by 25%. **1912.3.3.2**
Splices	Lap splices are not to be used: • Within joints. • Within a distance of twice the member depth from the face of the joint. • At locations where analysis indicates flexural yielding caused by inelastic lateral displacements of the frame. **1921.3.2.3**	No similar requirement.
Confinement Reinforcement	Hoops are required over a length equal to twice the member depth from the face of the supporting member toward midspan at both ends of the flexural member, with the first hoop located not more than 2 in. from the face of the support. **1921.3.3.1, 1921.3.3.2**	Zone 2 — Same requirement except that stirrups may be used rather than hoops. **1921.8.4.2** Zones 0 & 1 — No similar requirement.
Confinement Reinforcement	Hoops are required over lengths equal to twice the member depth on both sides of a section where flexural yielding may occur in connection with inelastic lateral displacements of the frame. **1921.3.3.1**	Reinforcement for flexural members subject to stress reversals shall consist of closed stirrups extending around flexural reinforcement. Also, provide minimum structural integrity reinforcement. **1907.11.2, 1907.13**
Confinement Reinforcement	Where hoops are required, the spacing shall not exceed: *d*/4 8 × diameter of smallest longitudinal bar 24 × diameter of hoop bars 12 in. **1921.3.3.2**	Zone 2 — Maximum stirrup spacing shall not exceed: *d*/4 8 × diameter of smallest longitudinal bar 24 × diameter of stirrup bars 12 in. **1921.8.4.2** Zones 0 & 1 — No similar requirement; provide required spacing of closed stirrups at location of stress reversals or torsion.
Confinement Reinforcement	Where hoops are required, longitudinal bars on the perimeter shall have lateral support conforming to 1907.10.5.3. **1921.3.3.3**	No similar requirement.
Confinement Reinforcement	Where hoops are not required, stirrups with seismic hooks at both ends shall be spaced at a distance not more than *d*/2 throughout the length of the member. **1921.3.3.4**	Zone 2 — Similar requirement except that seismic hooks are not required. **1921.8.4.3**
Confinement Reinforcement	Transverse reinforcment must also be proportioned to resist the design shear force (V_e). **1921.3.4**	Zone 2 — Transverse reinforcement must also be proportioned to resist the design shear force specified in 1921.8.3. **1921.8.3** Zones 0 & 1 — Provide sufficient transverse reinforcement for shear and torsion. **1911.5, 1911.6**

3.3.3 Splices

Lap splices must be placed at locations away from potential hinge areas subject to stress reversals under cyclic loading (see Fig. 3-2). Where lap splices are used, they should be designed as tension lap splices and must be properly confined. Welded splices and mechanical connections may also be used and must conform to 1921.2.6.

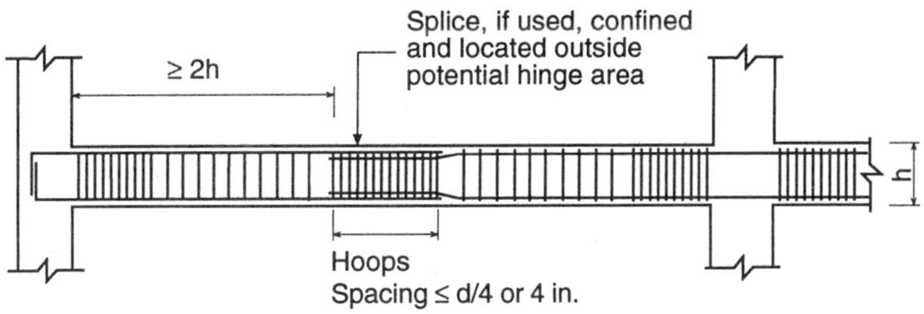

Typical Splice Details

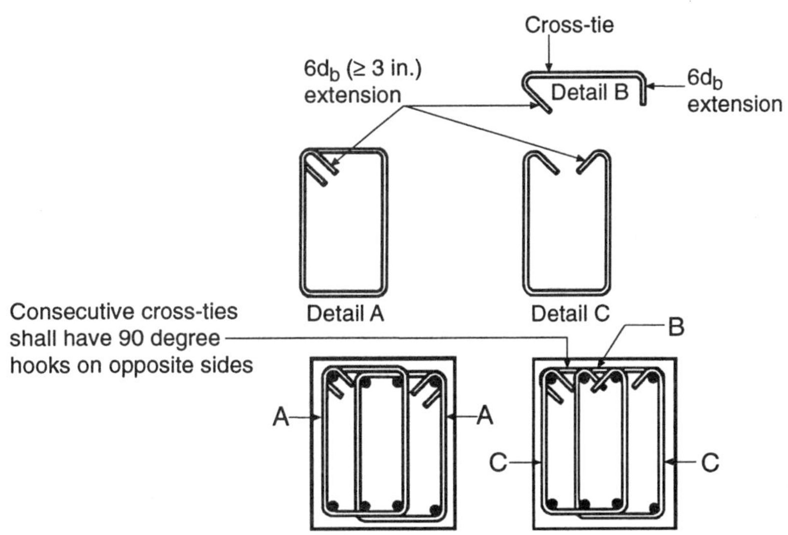

Single- and Two-Piece Hoops

Figure 3-2 Splices and Hoop Reinforcement in Seismic Zones 3 and 4

3.3.4 Confinement Reinforcement

Adequate confinement is required at ends of flexural members, where plastic hinges are likely to form, in order to ensure sufficient ductility of the members under reversible loads. Transverse reinforcement is also required at these locations to assist the concrete in resisting shear and to maintain lateral support for the reinforcing bars. In regions of high seismic risk (Zones 3 and 4), the transverse reinforcement for confinement must consist of hoops as shown in Fig. 3-3. Stirrups may be used for confinement in the case of Zone 2. Shear strength requirements are given in 1921.3.4 for Zone 3 and 4 and 1921.8.3 for Zone 2 (see Sect. 3.9).

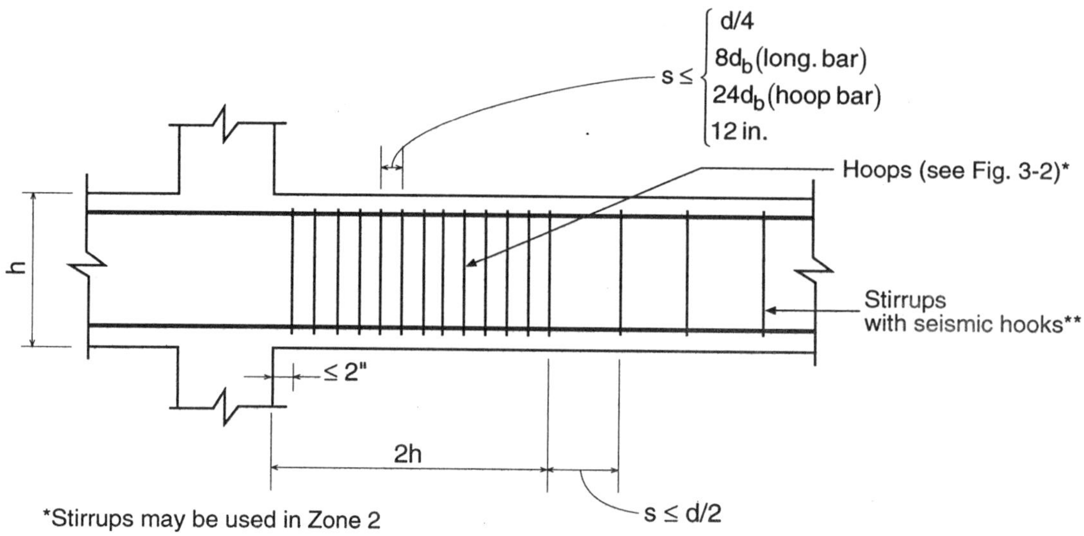

*Stirrups may be used in Zone 2
**Seismic hooks are not required in Zone 2

Figure 3-3 Transverse Reinforcement for Flexural Members in Seismic Zones 3 and 4

3.4 MEMBERS SUBJECTED TO BENDING AND AXIAL LOADS

Table 3-5 contains the requirements for frame members subjected to combined bending and axial loads. These requirements would typically apply to columns of frames and other flexural members which carry a factored axial load $P_u > A_g f_c' / 10$.

3.4.1 General

The geometric constraint requirements are based on results from previous practice and tests.

3.4.2 Flexural Requirements

Columns must be provided with sufficient strength so that they will not yield prior to the beams. Lateral sway caused by column hinging may result in excessive damage. Yielding of the columns prior to the beams could also result in total collapse of the structure. For these reasons, columns are designed with 20% higher flexural strength as compared to beams at the same joint, as shown in Fig. 3-4. The column flexural strength is calculated for the factored axial load, consistent with the direction of the lateral forces considered, resulting in the lowest flexural strength. The maximum allowable reinforcement ratio is reduced from 8% to 6% for Seismic Zones 3 and 4. This lower ratio prevents congestion of steel, which reduces the chance of improperly placed concrete. It also prevents the development of large shear stresses in the columns. Typically, providing a reinforcement ratio larger than about 4% is not practical or economical.

Table 3-5 Frame Members Subjected to Bending and Axial Loads

	Seismic Zones 3 & 4	Seismic Zones 0, 1 & 2
General	Frame members under this classification must meet the following requirements: • Factored axial compressive force > $A_g f'_c$ / 10. • Shortest cross-sectional dimension ≥ 12 in. • Ratio of shortest cross-sectional dimension to perpendicular dimension ≥ 0.4. **1921.4.1**	Zone 2 — Factored axial compressive force > $A_g f'_c$ / 10. **1921.8.2** Zones 0 & 1 — No similar requirements.
Flexural Requirements	The flexural strengths of the columns shall satisfy the following: $\Sigma M_e \geq (6/5)\ \Sigma M_g$ where ΣM_e = sum of moments at the center of the joint, corresponding to the design flexural strength of the columns. ΣM_g = sum of moments at the center of the joint, corresponding to the design flexural strength of the girders. If this requirement is not satisfied, the lateral strength and stiffness of the column shall not be considered when determining the strength and stiffness of the lateral force resisting system; also, the column must have transverse reinforcement over its full height per 1921.4.4, and shall conform to 1921.7. **1921.4.2**	No similar requirements.
	The reinforcement ratio (ρ_g) shall not be less than 0.01 and shall not exceed 0.06. **1921.4.3.1**	The reinforcement ratio (ρ_g) shall not be less than 0.01 and shall not exceed 0.08. **1910.9**
Splices	Lap splices are permitted only within the center half of the member length, must be tension splices, and have transverse reinforcement over the full lap length. Welded splices and mechanical connections shall conform to 1921.2.6. **1921.4.3.2**	There is no restriction on the location of splices which are typically located just above the floor for ease of construction. However, for Zone 2, splices may be located within the center half of the member as in Zones 3 & 4. Splices can be either Class A or B and shall have transverse reinforcement where reversal of flexural stress is anticipated.
Transverse Reinforcement	The transverse reinforcement requirements discussed in the following four items on the next page need only be provided over a length (ℓ_o) from each joint face and on both sides of any section where flexural yielding may occur. The length (ℓ_o) shall not be less than: depth of member 1/6 clear span 18 in. **1921.4.4.4**	Zone 2 — The length (ℓ_o) is the same as for Zones 3 & 4, except the maximum spacing (s_o) in this length, with the first tie being located at (s_o/2) from the joint face, shall not exceed h/2 8 × smallest longitudinal bar diameter 24 × transverse bar diameter 12 in. **1921.8.5.1** Zones 0 & 1 — No requirements for spacing of transverse reinforcement at column ends.

— Transverse Reinforcement continued on next page —

A_c = area of core of spirally reinforced compression member measured to outside diameter of spiral
A_{ch} = cross-sectional area of member measured out-to-out of transverse reinforcement
A_g = gross area of section
f'_c = compressive strength of concrete
f_{yh} = yield stress of transverse reinforcement
h = smallest cross-sectional dimension of the member
h_c = cross-sectional dimension of column core measured center-to-center of the confining reinforcement
s = spacing of transverse reinforcement

Table 3-5 Frame Members Subjected to Bending and Axial Loads (cont'd)

	Seismic Zones 3 & 4	Seismic Zones 0, 1 & 2
Transverse Reinforcement (continued)	Ratio of spiral reinforcement (ρ_s) shall not be less than the value given by: $$\rho_s = 0.12\frac{f'_c}{f_{yh}} \geq 0.45\left(\frac{A_g}{A_c} - 1\right)\frac{f'_c}{f_{yh}}$$ **1921.4.4.1**	Ratio of spiral reinforcement (ρ_s) shall not be less than the value given by: $$\rho_s = 0.45\left(\frac{A_g}{A_c} - 1\right)\frac{f'_c}{f_{yh}}$$ and shall conform to the provisions in 1907.10.4. **1910.9.3**
	Total cross-sectional area of rectangular hoop reinforcement for confinement (A_{sh}) shall not be less than that given by the following two equations: $$A_{sh} = 0.3\left(sh_c f'_c / f_{yh}\right)\left[\left(A_g / A_{ch}\right) - 1\right]$$ $$A_{sh} = 0.09\left(sh_c f'_c / f_{yh}\right)$$ **1921.4.4.1**	Transverse reinforcement must be provided to satisfy both shear and lateral support requirements for longitudinal bars. **1907.10.5, 1911.1**
	Transverse reinforcement shall be spaced at distances not exceeding $h/4$ or 4 in. **1921.4.4.2**	No similar requirements.
	Cross-ties or legs of overlapping hoops shall not be spaced more than 14 in. on center in the direction perpendicular to the longitudinal axis of the member. Vertical bars shall not be farther than 6 in. clear from a laterally supported bar. **1921.4.4.3, 1907.10.5.3**	Vertical bars shall not be farther than 6 in. clear from a laterally supported bar. **1907.10.5.3**
	Where the transverse reinforcement as discussed above is no longer required, the remainder of the column shall contain spiral or hoop reinforcement spaced at distances not to exceed 6 × longitudinal bar diameter 6 in. **1921.4.4.6**	Zone 2 — Tie spacing shall not exceed twice the spacing (s_o) determined per 1921.8.5.1. **1921.8.5.4** Zones 0 & 1 — Similar requirement except that maximum spacing shall not exceed: 16 × longitudinal bar diameter 48 × transverse bar diameter smallest cross-sectional dimension **1907.10.5.2**
	Any section where the nominal strength (ϕP_n) of the column is less than the sum of shears (V_e) for all the beams framing into the column, which are calculated using the probable moment strengths, transverse reinforcement shall be provided as discussed in the above items. **1921.4.4.7** Transverse reinforcement must also be proportioned to resist the design shear force (V_e). **1921.4.5**	Zone 2 — Transverse reinforcement must also be proportioned to resist the design shear forces specified in 1921.8.3. Zones 0 & 1 — Provide sufficient transverse reinforcement for shear. **1911.5.4, 1911.5.6**
	Columns supporting reactions from discontinued stiff members, such as walls, shall have transverse reinforcement as specified in 1921.4.4.1 through 1921.4.4.3 over their full height, if the factored axial compressive force, including earthquake effects, exceeds ($A_g f'_c$ / 10). This transverse reinforcement shall extend into the discontinued member for at least the development length of the largest longitudinal reinforcement in the column in accordance with 1921.5.4. If the column terminates on a footing or mat, the transverse reinforcement shall extend at least 12 in. into the footing or mat. **1921.4.4.5**	Zones 0, 1 & 2 — No similar requirement.

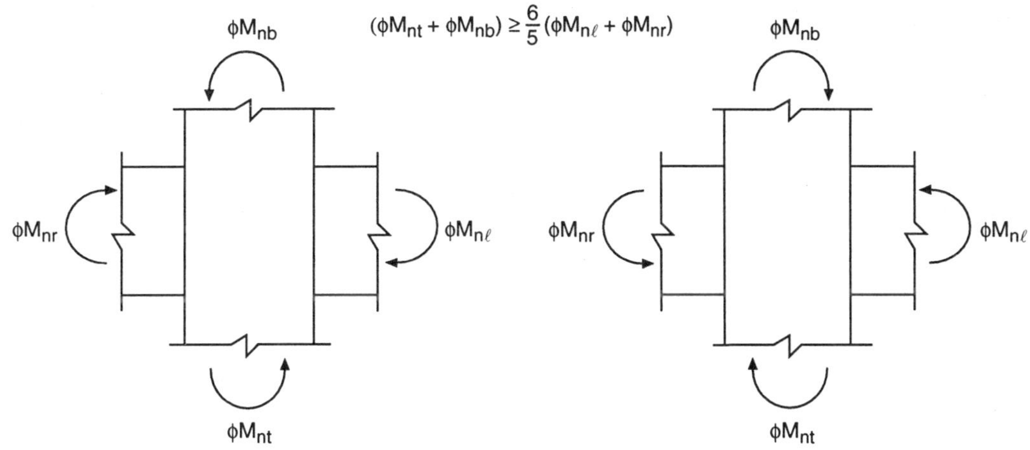

Subscripts ℓ, r, t, and b stand for left support, right support, top of column, and bottom of column, respectively.

Figure 3-4 "Strong Column-Weak Beam" Frame Requirements in Seismic Zones 3 and 4

3.4.3 Splices

Lap splices are to be located at midheight to prevent the splices from being in areas of spalling and high stress as shown in Fig. 3-5. The lap splices should be designed as tension lap splices and adequately confined with transverse reinforcement. Welded splices and mechanical connections must conform to 1921.2.6.

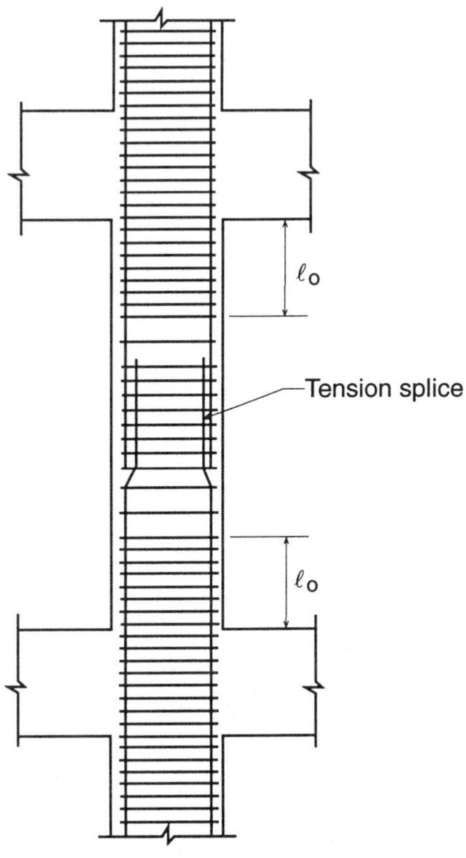

Figure 3-5 Typical Splice Details of Columns in Seismic Zones 3 and 4

3.4.4 Transverse Reinforcement

Column ends require adequate confinement to ensure column ductility in the event of hinge formation. They also require adequate shear reinforcement in order to prevent shear failures prior to the development of the flexural capacity of the section. The correct amount, spacing, and location of the transverse reinforcement must be provided so that both the confinement and shear strength requirements are satisfied. For Seismic Zones 3 and 4, the transverse reinforcement must be spiral or circular hoop reinforcement or rectangular hoop reinforcement, as shown in Fig. 3-6. For columns supporting discontinued stiff members (such as walls) as shown in Fig. 3-7, the transverse reinforcement needs to be provided over the full height of the column and must be extended at least the development length of the longitudinal bars in the column into the discontinued member (wall). The transverse reinforcement must also extend at least 12 in. into the footing or mat, if the column terminates on a footing or mat.

It is important to note that for columns where the point of contraflexure is not within the middle half of the clear height, transverse reinforcement as required in 1921.4.4.1 must be provided over the full height of the member.

Similar provisions are required for Zone 2 (see Fig. 3-6).

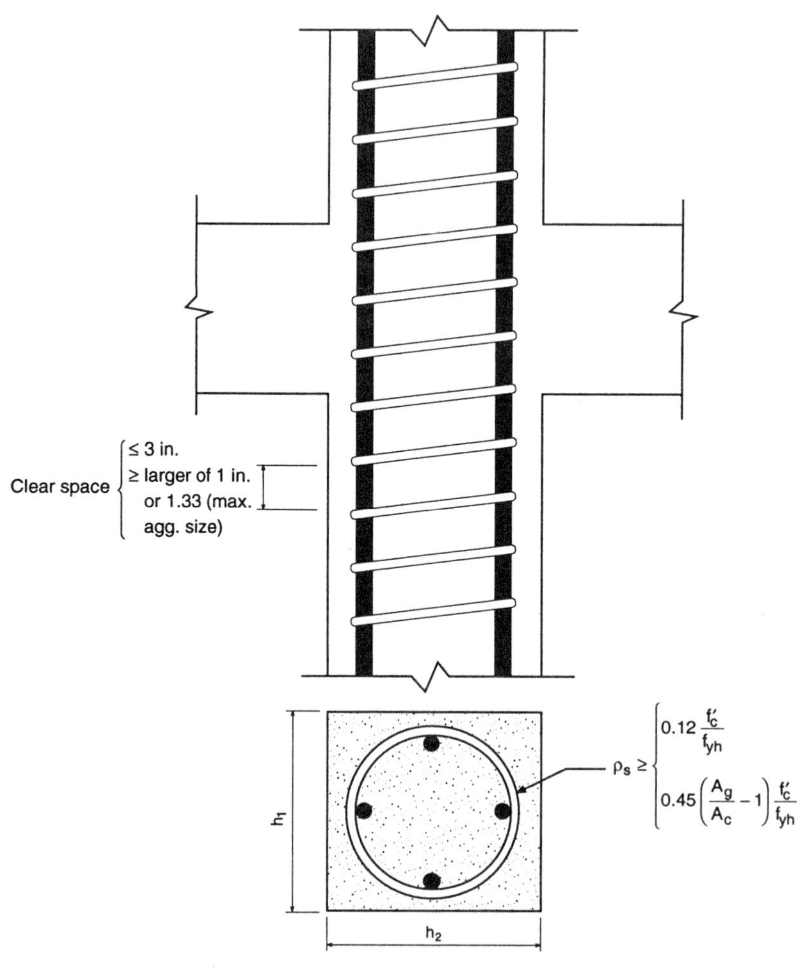

Figure 3-6 Confinement Requirements at Column Ends
(a) spiral hoop reinforcement

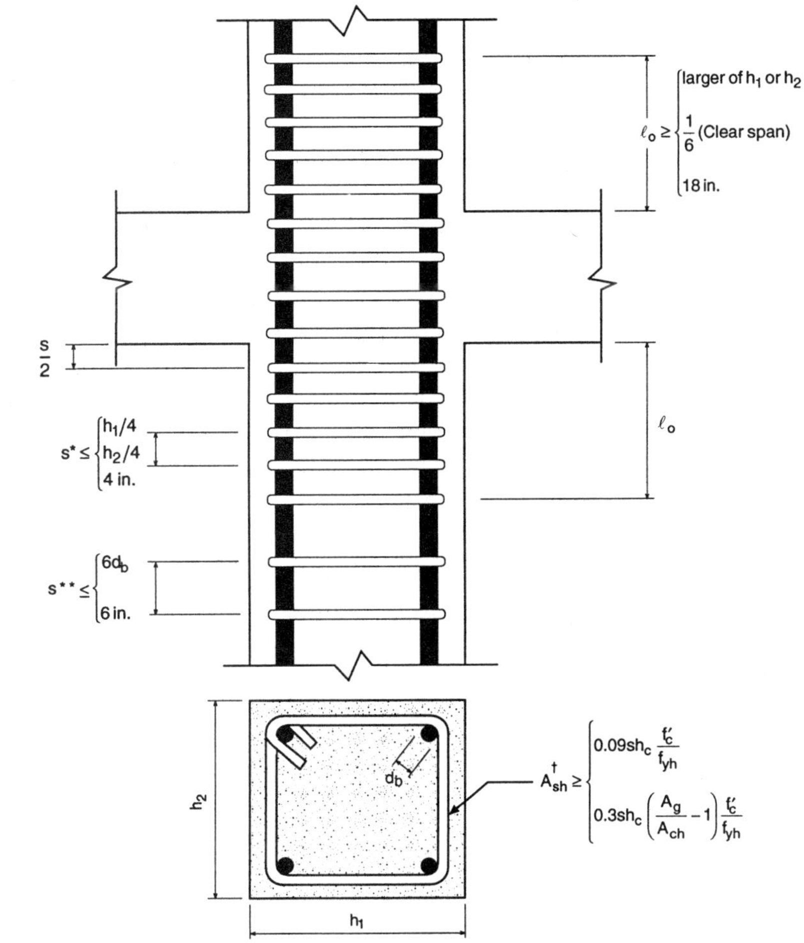

(b)

*For Zone 2, $s \leq$ $\frac{1}{2} \times$ smaller of h_1 or h_2; $8 \times$ smallest long. bar diameter; $24 \times$ transverse bar diameter; 12 in.

**For Zone 2, $s \leq 2 \times$ spacing required within ℓ_o

†Confinement requirements for Zones 3 & 4 need not be satisfied in Zone 2

Figure 3-6 Confinement Requirements at Column Ends
(b) rectangular hoop reinforcement

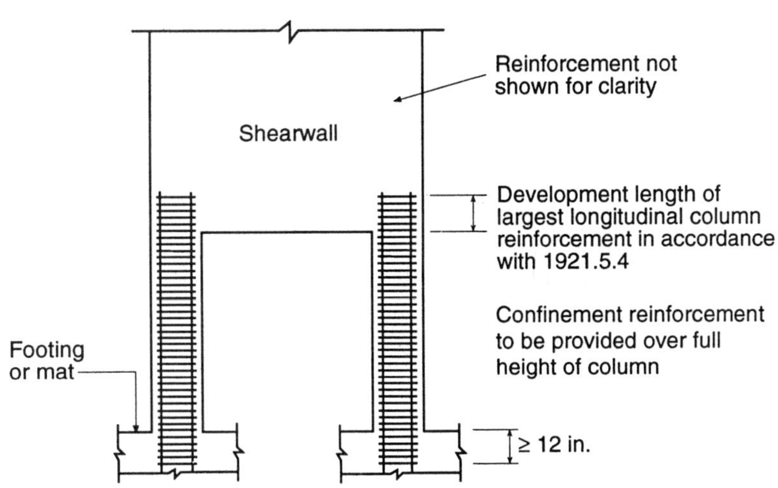

Figure 3-7 Columns Supporting Discontinued Stiff Members

3.5 SHEARWALLS

Table 3-6 contains the requirements for shearwalls. Special requirements exist for Seismic Zones 3 and 4 (1921.6). Walls in low to moderate seismic zones (0, 1 and 2) can be designed and detailed by the non-seismic provisions of the code.

3.5.1 Reinforcement

In Seismic Zones 3 and 4, structural walls are to be provided with reinforcement in two orthogonal directions in the plane of the wall (see Fig. 3-8). The minimum reinforcement ratio for both the longitudinal and transverse reinforcement is 0.0025, unless the design shear force does not exceed $A_{cv}\sqrt{f_c'}$, where A_{cv} is the net area of concrete bounded by the web thickness and the length of the wall in the direction of analysis; in this case, the minimum reinforcement must not be less than that given in 1914.3. The reinforcement provided for shear strength must be continuous and distributed uniformly across the shear plane with a maximum spacing of 18 in. At least two curtains of reinforcement are required if the in-plane factored shear force assigned to the wall exceeds $2A_{cv}\sqrt{f_c'}$. This serves to reduce fragmentation and premature deterioration of the concrete under load reversals into the inelastic range. Uniform distribution of reinforcement across the height and horizontal length of the wall helps control the width of the inclined (diagonal) cracks. Note that vertical reinforcement in the boundary elements of the wall is not to be included when checking the minimum reinforcement requirements.

3.5.2 Boundary Elements

Shearwall boundary zone detail requirements as prescribed in 1921.6.6.6 need not be provided in shearwalls or portions of shearwalls when

1. $P_u \le 0.10 A_g f_c'$ for geometrically symmetrical wall sections

   $\le 0.05 A_g f_c'$ for geometrically unsymmetrical wall sections

and either

2. $\dfrac{M_u}{V_u \ell_w} \le 1.0$

or

3. $V_u \le 3A_{cv}\sqrt{f_c'}$ and $\dfrac{M_u}{V_u \ell_w} \le 3$

For shearwalls and portions of shearwalls that do not meet these conditions and have a factored axial force $P_u < 0.35P_o$ (where P_o is the nominal axial load strength at zero eccentricity), boundary zones must be provided at each end a distance varying linearly from $0.25\ell_w$ to $0.15\ell_w$ for P_u between 0.35 P_o and 0.15 P_o. The boundary zone shall have a minimum length of $0.15\ell_w$ and shall be detailed in accordance with 1921.6.6.6.

Table 3-6 Shearwalls

	Seismic Zones 3 & 4	Seismic Zones 0, 1 & 2
Reinforcement	Reinforcement ratio shall not be less than 0.0025 along the longitudinal and transverse axes. If the design shear force $V_u \leq A_{cv}\sqrt{f'_c}$, provide minimum reinforcement of 1914.3. **1921.6.2.1**	Minimum vertical reinforcement ratio = 0.0012 for No. 5 bars or smaller = 0.0015 for No. 6 bars or larger Minimum horizontal reinforcement ratio = 0.0020 for No. 5 bars or smaller = 0.0025 for No. 6 bars or larger **1914.3**
	At least two curtains of reinforcement shall be used in a wall if the in-plane factored shear force (V_u) assigned to the wall exceeds $2\,A_{cv}\sqrt{f'_c}$. **1921.6.2.2**	Walls more than 10 in. thick require two curtains of reinforcement (except basement walls). **1914.3.4**
	Reinforcement spacing each way shall not exceed 18 in. **1921.6.2.1**	Reinforcement spacing shall not exceed: 3 × wall thickness 18 in. **1914.3.5**
	All continuous reinforcement in structural walls, diaphragms, trusses, struts, ties, chords, and collector elements shall be anchored or spliced in accordance with the provisions for reinforcement in tension as specified in 1921.5.4. **1921.6.2.4**	The development lengths, spacing, and anchorage of reinforcement shall be as per 1912, 1914, and 1915.
	Except when V_u in the plane of the wall is less than $A_{cv}\sqrt{f'_c}$, transverse reinforcement terminating at the edges of structural walls without boundary elements shall have a standard hook engaging the edge reinforcement, or the edge reinforcement shall be enclosed in U-stirrups having the same size and spacing as, and spliced to, the transverse reinforcement. Welded splices and mechanical connections shall conform to 1921.2.6. **1921.6.2.2**	No similar requirement.

A_{cv} = net area of concrete section bounded by web thickness and length of section in the direction of shear force considered

A_g = gross area of section

b_w = width of web

d = effective depth of section

f'_c = compressive strength of concrete

f_{yh} = yield strength of transverse reinforcement

h_c = cross-sectional dimension of shearwall boundary zone measured center-to-center of confining reinforcement

ℓ_n = clear span of coupling beam

ℓ_u = unsupported length of wall

ℓ_w = length of entire wall or segment of wall in direction of shear force

s = spacing of transverse reinforcement

Table 3-6 Shearwalls (cont'd)

	Seismic Zones 3 & 4	Seismic Zones 0, 1 & 2
Boundary Elements	Boundary zone details given in 1921.6.6.6 shall be provided when the conditions specified in 1921.6.6.4 are not satisfied. Alternatively, the boundary zone details shall be provided over those portions of the wall where compressive strains exceed 0.003. **1921.6.6.4 and 1921.6.6.5**	No similar requirement.
	Dimensional requirements. 1. All portions of the boundary zones shall have a thickness of $\ell_u/16$ or greater. 2. Boundary zones shall extend vertically a distance equal to the development length of the largest vertical bar within the boundary zone above the elevation where the requirements of 1921.6.6.4 or 1921.6.6.5 are met. Extensions below the base of the boundary zone shall conform to 1921.4.4.6. Exception: The boundary zone reinforcement need not extend above the base of the boundary zone a distance greater than the larger of ℓ_w or $M_u/4V_u$. 3. Boundary zones as determined by the requirements of 1921.6.6.5 shall have a minimum length of 18 inches at each end of the wall or portion of wall. 4. In I-, L-, C- or T-shaped sections, the boundary zone at each end shall include the effective flange width and shall extend at least 12 inches into the web. **Confinement reinforcement.** 1. All vertical reinforcement within the boundary zone shall be confined by hoops or cross-ties producing an area of steel not less than: $A_{sh} = 0.09sh_cf'_c/f_{yh}$ 2. Hoops and cross-ties shall have a vertical spacing not greater than the smaller of 6 inches or 6 diameters of the largest vertical bar within the boundary zone. 3. The ratio of the length to the width of the hoops shall not exceed 3. All adjacent hoops shall be overlapping. 4. Cross-ties or legs of overlapping hoops shall not be spaced further apart than 12 inches along the wall. 5. Alternate vertical bars shall be confined by the corner of a hoop or cross-tie. **Horizontal reinforcement.** 1. All horizontal reinforcement terminating within a boundary zone shall be anchored in accordance with 1921.6.2. 2. Horizontal reinforcement shall not be lap spliced within the boundary zone. **1921.6.6.6**	No similar requirement.

— Boundary Elements continued on next page —

Table 3-6 Shearwalls (cont'd)

	Seismic Zones 3 & 4	Seismic Zones 0, 1 & 2
Boundary Elements (cont'd)	**Vertical reinforcement.** 1. Vertical reinforcement shall be provided to satisfy all tension and compression requirements. 2. Area of reinforcement shall not be less than 0.005 times the area of boundary zone or less than two No. 5 bars at each edge of boundary zone. 3. Lap splices of vertical reinforcement within the boundary zone shall be confined by hoops or cross-ties. Spacing of hoops and cross-ties confining lap-spliced reinforcement shall not exceed 4 inches. **1921.6.6.6**	No similar requirement.
Coupling Beams	For coupling beams with $\ell_n/d \geq 4$, the design shall conform to the requirements of 1921.2 and 1921.3. It shall be permitted to waive the requirements of 1921.3.1.3 and 1921.3.1.4 if it can be shown by analysis that lateral stability is adequate or if alternative means of maintaining lateral stability is provided. **1921.6.10.1** Coupling beams with $\ell_n/d < 4$ shall be permitted to be reinforced with two intersecting groups of symmetrical diagonal bars. Coupling beams with $\ell_n/d < 4$ and with factored shear force (V_u) exceeding $4\sqrt{f'_c}\, b_w d$ shall be reinforced with two intersecting groups of symmetrical diagonal bars. Each group shall consist of a minimum of four bars assembled in a core with a lateral dimension of each side not less than $b_w/2$ or 4 inches. The design shear strength (ϕV_n) of these coupling beams shall be determined by: $\phi V_n = 2\phi\, f_y \sin\alpha\, A_{vd} \leq 10\phi\sqrt{f'_c}\, b_w d$ where: α = the angle between the diagonal reinforcement and the longitudinal axis A_{vd} = the total area of reinforcement of each group of diagonal bars ϕ = 0.85 **1921.6.10.2** Each group of diagonally placed bars shall be enclosed in transverse reinforcement conforming to 1921.4.4.1 through 1921.4.4.3. For purpose of computing A_g, as per Eqs. (10-6) and (21-3), the minimum cover, as specified in 1907.7, shall be assumed over each group of diagonally placed reinforcing bars. **1921.6.10.3** Reinforcement parallel and transverse to the longitudinal axis shall be provided and, as a minimum, shall conform to 1910.5, 1911.8.9, and 1911.8.10. **1921.6.10.4** Contribution of the diagonal reinforcement to nominal flexural strength of the coupling beam shall be considered. **1921.6.10.5**	No similar requirement.

Table 3-6 Shearwalls (cont'd)

	Seismic Zones 3 & 4	Seismic Zones 0, 1 & 2
Wall Piers	Wall piers not designed as part of a special moment-resisting frame shall have transverse reinforcement designed to satisfy the requirements of 1921.6.13.2, unless the piers satisfy 1921.7. Also the requirements of 1921.6.13.2 need not be satisfied for wall piers along a wall line within a story where other shear wall segments provide lateral support to the wall piers, and such segments have a total stiffness of at least six times the sum of the stiffnesses of all the wall piers. **1921.6.13.1** Transverse reinforcement shall be designed to resist the shear forces determined from 1921.4.5.1 and 1921.3.4.2. When the axial compressive force, including earthquake effects, is less than $A_g f'_c/20$, transverse reinforcement in wall piers may have standard hooks at each end in lieu of hoops. Spacing of transverse reinforcement shall not exceed 6 inches. Transverse reinforcement shall be extended beyond the pier clear height for at least the development length of the largest longitudinal reinforcement in the wall pier. **1921.6.13.2** Wall segments with a horizontal length-to-thickness ratio less than 2.5 shall be designed as columns. **1921.6.13.3**	No similar requirement.

Alternatively, the requirements for boundary zones may be based on the compressive strain level at the wall edges when the wall is subjected to displacement levels resulting from ground motion specified in 1629.2 using cracked section properties. The boundary zone details defined in 1921.6.6.6 shall be provided over those portions of the wall where the compressive strain exceeds 0.003. In no case shall the compressive strain exceed 0.015. Figure 3-9 shows two typical layouts for boundary elements. Detailing requirements are given in Table 3-6.

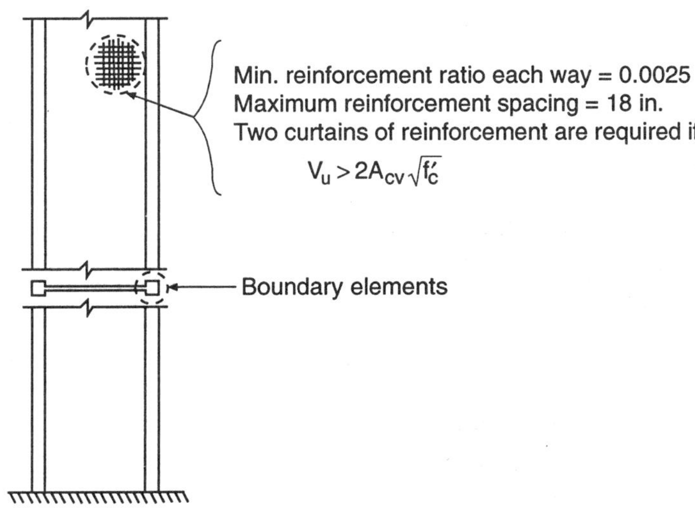

Figure 3-8 Shearwall Design and Detailing Requirements

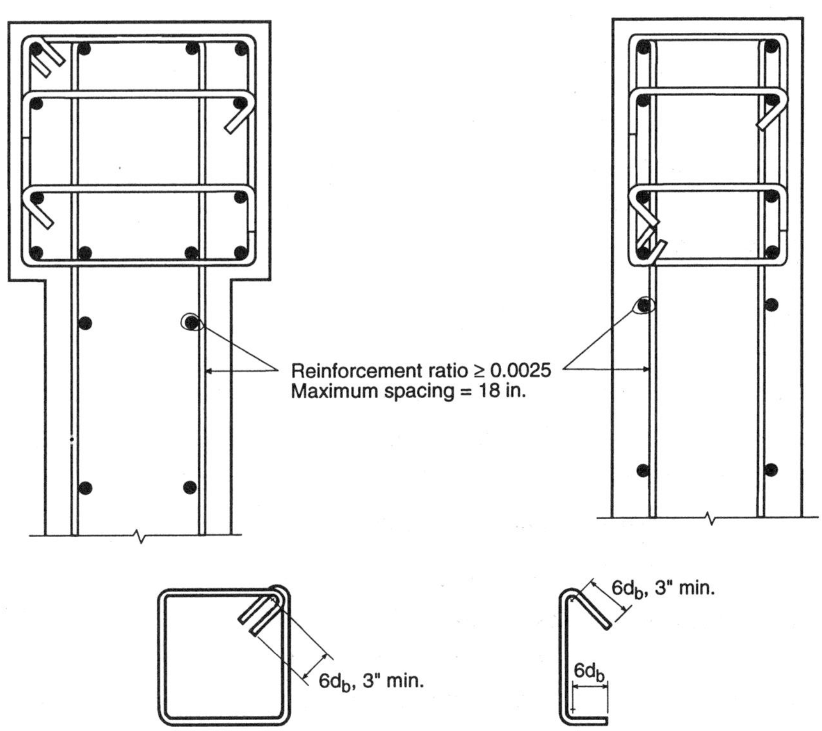

Figure 3-9 Boundary Element Layout

3.5.3 Coupling Beams

When adequately proportioned and detailed, coupling beams between walls can provide an efficient means of energy dissipation under seismic forces. Due to their relatively short length, ends of coupling beams are usually subjected to large inelastic rotations. Adequate detailing and shear reinforcement are necessary to prevent shear failure and ensure ductility and energy dissipation. For these reasons, confined diagonal reinforcing bars are required; the ends of these bars must be adequately anchored in the shearwalls.

3.6 DIAPHRAGMS

The requirements for diaphragms are similar to those for shearwalls. The minimum reinforcement ratio for structural diaphragms is given in 1907.12, with a maximum bar spacing of 18 in. (1921.6.2.1).

The general provisions for the design of diaphragms is given in 1921.6.4. Boundary element requirements are given in 1921.6.7.

3.7 JOINTS OF FRAMES

The overall integrity of a structure is dependent on the behavior of the beam-column joint. Degradation of the joint can result in large lateral deformations which can cause excessive damage or even failure. Table 3-7 contains the requirements for frame joints. For Seismic Zones 0, 1 and 2, the beam-column joints do not require the special design and detailing requirements as for Zones 3 and 4. Although special detailing is not required for Zone 2, it may be prudent to provide the same details as required for Zones 3 and 4.

3.7.1 Longitudinal Reinforcement

Slippage of the longitudinal reinforcement in a beam-column joint can lead to an increase in the joint rotation. Longitudinal bars must be continued through the joint or must be properly developed for tension (1921.5.4) and compression (1912) in the confined column core. The minimum column size requirement of 1921.5.1.4 reduces the possibility of failure from loss of bond during load reversals beyond the yield point of the steel.

3.7.2 Shear Strength

The most significant factor in determining the capacity of a beam-column joint is the effective area (A_j) of the joint, as shown in Fig. 3-10. For joints that are confined by beams on all four faces, the shear capacity of the joint is equal to $20\sqrt{f'_c}A_j$. If the joint is confined only on three faces, or on two opposite faces, the capacity must be reduced by 25% to $15\sqrt{f'_c}A_j$. For other cases, the shear capacity is equal to $12\sqrt{f'_c}A_j$. It is important to note that the shear capacity is a function of the concrete strength and the cross-sectional area only. Test results show that the shear strength of the joint is not altered significantly with changes in transverse reinforcement, provided a minimum amount of such reinforcement is present. Thus, only the concrete strength or the member size can be modified if the shear capacity of the beam-column joint is inadequate.

Table 3-7 Joints of Frames

	Seismic Zones 3 & 4	Seismic Zones 0, 1 & 2
Longitudinal Beam Reinforcement	Beam longitudinal reinforcement terminated in a column shall be extended to the far face of the confined column core and anchored in tension according to 1921.5.4 and in compression according to 1912. **1921.5.1.3**	No similar requirement.
	Where longitudinal beam reinforcement extends through a joint, the column dimension parallel to the beam reinforcement shall not be less than 20 times the diameter of the largest longitudinal bar for normal weight concrete. For lightweight aggregate concrete, this dimension shall be not less than 26 times the bar diameter. **1921.5.1.4**	No similar requirements.
Shear Strength	The nominal shear strength of the joint shall not exceed the forces specified below for normal weight aggregate concrete. For joints confined on all four faces........ $20\sqrt{f'_c}\,A_j$ For joints confined on three faces or on two opposite faces .... $15\sqrt{f'_c}\,A_j$ For other joints: .... $12\sqrt{f'_c}\,A_j$ where: A_j = effective cross-sectional area within a joint in a plane parallel to the plane of the reinforcement generating shear in the joint. The joint depth shall be the overall depth of the column. Where a beam frames into a support of larger width, the effective width of the joint shall not exceed the smaller of: 1. Beam width plus the joint depth. 2. Twice the smaller perpendicular distance from the longitudinal axis of the beam to the column side. A joint is considered to be confined if confining members frame into all faces of the joint. A member that frames into a face is considered to provide confinement at the joint if at least 3/4 of the face of the joint is covered by the framing member. **1921.5.3**	Although it is not required, it may be prudent to check the shear strength of the joint in Zone 2. The force in the longitudinal beam reinforcement may be taken as $1.0f_y$ rather than the $1.25f_y$ required for Zones 3 & 4.

f'_c = compressive strength of concrete

f_y = yield strength of reinforcement

Table 3-7 Joints of Frames (cont'd)

	Seismic Zones 3 & 4	Seismic Zones 0, 1 & 2
Shear Strength (continued)	In determining shear forces in the joints, forces in the longitudinal beam reinforcement at the joint face shall be calculated by assuming that the stress in the flexural tensile reinforcement is $1.25f_y$. **1921.5.1.1**	No similar requirement.
	For lightweight aggregate concrete, the nominal shear strength of the joint shall not exceed 3/4 of the limits given in 1921.5.3.1. **1921.5.3.2**	No similar requirement.
Transverse Reinforcement	The transverse hoop reinforcement required for column ends (1921.4.4) shall be provided within the joint, unless the joint is confined by structural members as specified in 1921.5.2.2. If members frame into all four sides of the joint and the member width at the column face is at least 3/4 the column width, the transverse reinforcement can be reduced to 50% of the requirements of 1921.4.4.1 within the depth of the shallowest member. The spacing shall not exceed 6 in. at these locations. **1921.5.2.1, 1921.5.2.2**	In Zone 2, it may be prudent to continue the column end transverse reinforcement through the joint as in the case of Zones 3 & 4. Zones 0 & 1 — No similar requirement.

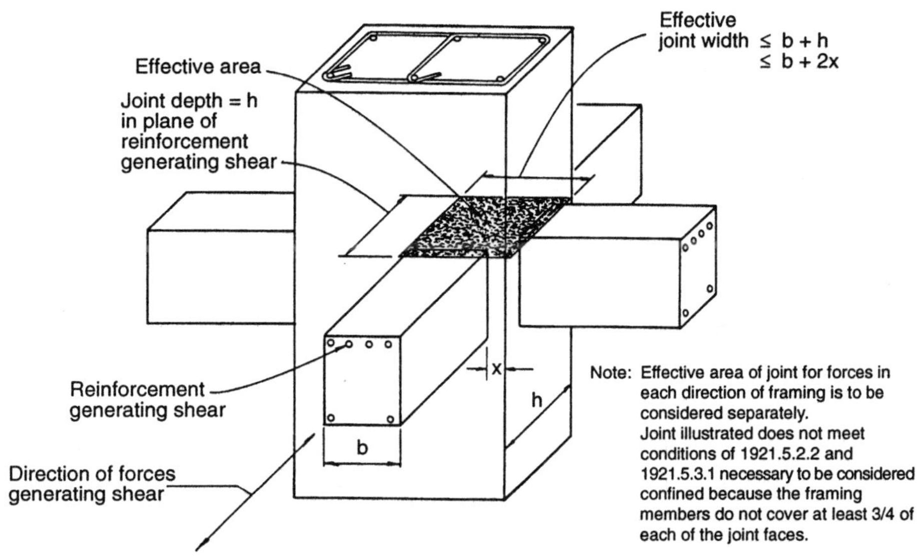

Figure 3-10 Effective Area of Joint (A_j)

The larger the tension force in the steel, the greater the shear in the joint (Fig. 3-11). Thus, the tensile force in the reinforcement is conservatively taken as $1.25 f_y A_s$. The multiplier of 1.25 takes into account the likelihood that due to strain hardening and actual strengths higher than the specified yield strengths, a larger tensile force may develop in the bars, resulting in a larger shear force.

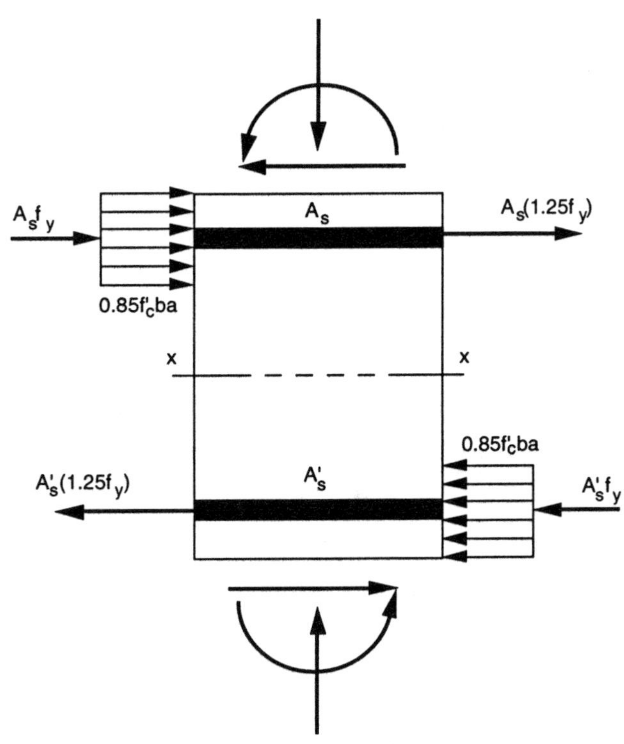

Figure 3-11 Horizontal Shear in Beam-Column Connection

3.7.3 Transverse Reinforcement

The transverse reinforcement in a beam-column connection is intended to provide adequate confinement of the concrete to ensure its ductile behavior and to allow it to maintain its vertical load-carrying capacity even after spalling of the outer shell.

Minimum confinement reinforcement of the same amount required for potential hinging regions in columns, as defined by Eqs. (10-6), (21-2), (21-3), or (21-4), must be provided within a beam-column joint around the column reinforcement, unless the joint is confined by structural members as specified in 1921.5.2.2.

For joints confined on all four faces, a 50% reduction in the amount of confinement reinforcement is allowed. A member that frames into a face is considered to provide confinement to the joint if at least three-quarters of the face of the joint is covered by the framing member. The code further allows that where a 50% reduction in the amount of confinement reinforcement is permissible, the spacing specified in 1921.4.4.2(b) may be increased to 6 in. (1921.5.2.2).

The minimum amount of confinement reinforcement, as noted above, must be provided through the joint regardless of the magnitude of the calculated shear force in the joint. The 50% reduction in the amount of confinement reinforcement allowed for joints having horizontal members framing into all four sides recognizes the beneficial effect provided by these members in resisting the bursting pressures that can be generated within the joint.

3.8 Development Lengths

Table 3-8 contains the requirements for development lengths of longitudinal reinforcement. Provisions are given for both hooked and straight bars.

3.8.1 Hooked Bars

A standard 90-degree hook is depicted in Fig. 3-12. The minimum development lengths are practically the same for all seismic zones. Equation (21-5), based on the requirements of 1912.5, includes the factors for hooks enclosed in ties (0.8), satisfaction of minimum cover requirements (0.7), a cyclic load factor (1.1), and a factor of 1.25 for overstrength in the reinforcing steel. The equation in 1912.5.2 $\left(1{,}200 d_b / \sqrt{f_c'}\right)$ is multiplied by these factors to obtain the equation that is given in 1921.5.4.1 for reinforcing bars with f_y = 60,000 psi:

$$\ell_{dh} = \frac{0.8 \times 0.7 \times 1.25 \times 1.1 \times 1{,}200 d_b}{\sqrt{f_c'}}$$

$$= \frac{924 d_b}{\sqrt{f_c'}} = \frac{f_y d_b}{65\sqrt{f_c'}}$$

Table 3-8 Development Lengths

	Seismic Zone 3 & 4	Seismic Zones 0, 1 & 2
Hooked Bars	The development length (ℓ_{dh}) for a bar with a standard 90° hook in normal weight aggregate concrete located within a confined core of a column or a boundary member shall not be less than: $8d_b$ 6 in. $f_y\, d_b / \left(65\sqrt{f'_c}\right)$ **1921.5.4.1**	The development length for a hooked bar in normal weight aggregate concrete must not be less than: $8d_b$ 6 in. $1{,}200 d_b/\sqrt{f'_c}$ for $f_y = 60{,}000$ psi However, the following reductions are allowed: • For No. 11 bars and smaller, side cover not less than 2.5 in. and with cover on the bar extension beyond hook of not less than 2 in., a reduction multiplier of 0.7 is permitted. • For No. 11 bars and smaller, hooks enclosed in ties spaced along the full development length not greater than $3d_b$, a reduction multiplier of 0.8 is permitted. • Where anchorage or development for f_y is not specifically required, a multiplier of (A_s required)/(A_s provided) can be used. **1912.5**
	For lightweight aggregate concrete, the development length for the hooked bars shall be increased by 25%. **1921.5.4.1**	If lightweight aggregate is used, the development length for the hooked bars shall be increased by 30%. For epoxy-coated reinforcement, development length shall be increased by 20%. **1912.5.3.5, 1912.5.3.6**

A_s = area of tension reinforcement

A_{tr} = cross-sectional area of transverse reinforcement within the spacing (s) which crosses the potential plane of splitting through the reinforcement being developed

d_b = diameter of bar

f'_c = compressive strength of concrete

f_{ct} = average splitting tensile strength of lightweight aggregate concrete

f_y = yield strength of reinforcement

f_{yt} = yield strength of transverse reinforcement

n = number of bars being developed

s = maximum spacing of transverse reinforcement

Table 3-8 Development Lengths (cont'd)

	Seismic Zones 3 & 4	Seismic Zones 0, 1 & 2
Straight Bars	For straight bars (No. 11 or smaller), the development length shall be 2.5 times the length required for hooked bars. If the bar is a top bar (depth of concrete cast in one lift beneath the bar exceeds 12 in.), the length shall be increased by 3.5 times the length required for hooked bars. **1921.5.4.2** Straight bars terminated at a joint shall pass through the confined core of a column or of a boundary member. Any portion of the straight embedment length not within the confined core shall be increased by a factor of 1.6. **1921.5.4.3** If epoxy-coated reinforcement is used, the above development lengths shall be multiplied by the applicable factors in 1912.2.4 or 1912.5.3.6. **1921.5.4.4**	For straight bars, the development length (ℓ_d) can be determined from the following: $$\frac{\ell_d}{d_b} = \frac{3}{40} \frac{f_y}{\sqrt{f'_c}} \frac{\alpha\beta\gamma\lambda}{\left(\frac{c + K_{tr}}{d_b}\right)}$$ in which the term $(c + K_{tr})/d_b$ shall not be taken greater than 2.5. ***α = reinforcement location factor*** Top bar .... 1.3 Other reinforcement .... 1.0 ***β = coating factor*** Epoxy-coated (with cover < $3d_b$, or clear spacing < $6d_b$) .... 1.5 All other epoxy-coated bars or wires .... 1.2 Uncoated reinforcement .... 1.0 The product of $\alpha\beta$ need not be taken greater than 1.7. ***γ = reinforcement size factor*** No. 6 and smaller bars and deformed wires .... 0.8 No. 7 and larger bars .... 1.0 ***λ = lightweight aggregate concrete factor*** Lightweight aggregate concrete .... 1.3 When f_{ct} is specified, $\lambda = 6.7\sqrt{f'_c}/f_{ct} \geq 1.0$ Normal weight concrete .... 1.0 ***c = spacing or cover dimension*** ***K_{tr} = transverse reinforcement index*** $= \frac{A_{tr} f_{yt}}{1500sn}$ **1912.2.3**†

†*Provisions of 1912.2.2 may also be used.*

Therefore, with the exceptions for the overstrength of steel yield factor and the cyclic loading factor used for Zones 3 and 4, the development length for No. 11 bars and smaller is identical in all seismic zones for hooked bars which satisfy the minimum concrete cover and transverse reinforcement requirements.

When hooked bars are being developed in lightweight aggregate concrete, the length must be increased by 25% for Zones 3 and 4. This is slightly less than the 30% required in other zones.

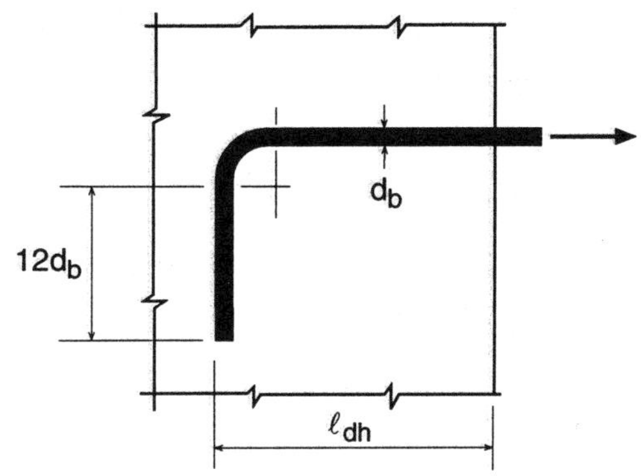

Figure 3-12 Standard 90-Degree Hook

3.8.2 Straight Bars

In Seismic Zones 3 and 4, the development length for No. 11 straight bars and smaller is determined by multiplying the development length for hooked bars by (a) two-and-a-half (2.5) times the length required by 1921.5.4.1 if the depth of the concrete cast in one lift beneath the bar does not exceed 12 in., and (b) three-and-a-half (3.5) times the length required by 1921.5.4.1 if the depth of the concrete cast in one lift beneath the bar exceeds 12 in. If a portion of a straight bar is not located within the confined core of a column or boundary element, the length of that bar shall be increased by an additional 60%. Provisions for epoxy-coated bars are given in 1921.5.4.4.

3.9 Shear Strength Requirements

Table 3-9 contains the shear strength requirements for flexural members, members subjected to combined axial and flexural loads, shearwalls, and diaphragms. It is important to note that the shear reinforcement requirement must be compared with the confinement requirement wherever applicable, so that the transverse reinforcement which is provided satisfies both criteria.

3.9.1 Flexural Members

Typically, larger forces than those prescribed by the codes are induced in structural members during an actual earthquake. Designing for shear forces using the code-prescribed load combinations is not conservative since the reinforcement may go beyond its yield strength resulting in larger than anticipated shear forces. Thus, the flexural members should be designed for shear using the maximum probable moment strength (M_{pr}) (which is the capacity of the beam with the stress in the reinforcing steel $f_s = 1.25f_y$ and a strength reduction factor of 1.0) and the tributary gravity load along its span (see Fig. 3-13). Note that sidesway to the right and to the left must both be considered to obtain the maximum V_e.

Table 3-9 Shear Strength Requirements

	Seismic Zones 3 & 4	Seismic Zones 0, 1 & 2
Flexural Members	The design shear force (V_e) shall be determined from consideration of the static forces on the portion of the member between the faces of the joints. It shall be assumed that the moments at the beam ends are equal to the probable strength moment (M_{pr}) and that the member is loaded with the tributary gravity load along the span. The probable strength moment is calculated as the moment capacity of the beam with $f_s = 1.25\ f_y$ and a strength reduction factor of 1.0. **1921.3.4.1**	Zone 2 — Two options are available for determining shear strength in beams, columns, and two-way slabs: Option 1 — Same as for Zones 3 & 4, except that the nominal moment strength (M_n) is used rather than the probable strength moment (M_{pr}). Option 2 — Use maximum shear obtained from design load combinations with the earthquake effects (E) doubled. **1921.8.3** Zones 0 & 1 — Designed for actual factored forces.
Flexural Members	If the earthquake-induced shear force represents one-half or more of the maximum required shear strength, the contribution of the concrete to the shear carrying capacity shall be equal to zero if the factored axial compressive force including earthquake effects is less than $A_g f'_c\ /\ 20$. **1921.3.4.2**	No similar requirement.
Members Subjected to Bending and Axial Loads	Shear in frame members subjected to combined bending and axial load shall be determined using the maximum probable moment strength of the member associated with the range of factored axial loads on the member. The member shear need not exceed that determined from joint strengths, based on the probable moment strength of the transverse members framing into the joints. In no case shall the design shear force be less than the factored shear force determined from analysis of the structure. **1921.4.5.1**	The shear in frame members subjected to combined bending and axial force are determined in the same manner as for flexural members.

A_{cv} = net area of concrete section bounded by the web thickness and the length of section in the direction of shear force considered

A_g = gross area of section

A_{sv} = projection on A_{cv} of area of distributed shear reinforcement crossing the plane of A_{cv}

A_v = area of horizontal shear reinforcement within a distance s_2

d = distance from extreme compression fiber to centroid of longitudinal tension reinforcement = 0.8 ℓ_w for walls

f'_c = compressive strength of concrete

f_s = tensile stress in the reinforcement

f_y = yield strength of reinforcement

h = overall thickness of member

h_w = height of entire wall (diaphragm) or the segment of the wall (diaphragm) considered

ℓ_w = length of entire wall (diaphragm) or the segment of the wall (diaphragm) considered in the direction of the shear force

N_u = factored axial load normal to cross-section

V_c = nominal shear strength provided by concrete

V_n = nominal shear strength

V_s = nominal shear strength provided by shear reinforcement

ρ_h = ratio of horizontal shear reinforcement area to gross concrete area of vertical section

ρ_n = ratio of distributed shear reinforcement on a plane perpendicular to plane of A_{cv} (Zones 3 & 4)

= ratio of vertical shear reinforcement area to gross concrete area of horizontal section (Zones 0, 1 & 2)

$\rho_v = A_{sv} / A_{cv}$

Table 3-9 Shear Strength Requirements (cont'd)

	Seismic Zones 3 & 4	Seismic Zones 0, 1 & 2
Walls and Diaphragms	The nominal shear strength (V_n) for shearwalls and diaphragms when (h_w / ℓ_w) is 2 or greater is: $V_n = A_{cv}\left(2\sqrt{f'_c} + \rho_n f_y\right)$ When h_w / ℓ_w is less than 2: $V_n = A_{cv}\left(\alpha_c\sqrt{f'_c} + \rho_n f_y\right)$ α_c varies linearly from 3.0 for $h_w / \ell_w = 1.5$ to 2.0 for $h_w / \ell_w = 2.0$. **1921.6.5.2, 1921.6.5.3**	The nominal shear strength (V_n) for shearwalls can be calculated using the following methods: $V_c = 3.3\sqrt{f'_c}\,hd + \frac{N_u d}{4\ell_w}$ or $V_c = \left[0.6\sqrt{f'_c} + \frac{\ell_w\left(1.25\sqrt{f'_c} + 0.2\frac{N_u}{\ell_w h}\right)}{\frac{M_u}{V_u} - \frac{\ell_w}{2}}\right] hd$ where $\frac{M_u}{V_u} - \frac{\ell_w}{2} \geq 0$ $V_s = \frac{A_v f_y d}{s_2}$ $V_n = V_c + V_s$ **1911.10**
	Walls (diaphragms) shall have distributed shear reinforcement providing resistance in two orthogonal directions in the plane of the wall (diaphragm). If the ratio (h_w / ℓ_w) does not exceed 2.0, reinforcement ratio (ρ_v) shall not be less than reinforcement ratio (ρ_n). **1921.6.5.5**	Ratio (ρ_h) of horizontal shear reinforcement area to gross concrete area of vertical section shall not be less than 0.0025. **1911.10.9.2** Spacing of horizontal shear reinforcement (s_2) shall not exceed $\ell_w/5$, $3h$, nor 18 in. **1911.10.9.3** The minimum vertical reinforcement ratio (ρ_n) is a function of (h_w / ℓ_w) and of the horizontal reinforcement as shown below: $\rho_n = 0.0025 + 0.5\left(2.5 - \frac{h_w}{\ell_w}\right)(\rho_h - 0.0025) \geq 0.0025$ Also, the vertical shear reinforcement need not exceed the required horizontal reinforcement. **1911.10.9.4** Spacing of vertical shear reinforcement (s_1) shall not exceed $\ell_w/3$, $3h$, nor 18 in. **1911.10.9.5**
	Nominal shear strength of all wall piers sharing a common lateral force shall not be assumed to exceed $8A_{cv}\sqrt{f'_c}$, where A_{cv} is the total cross-sectional area, and the nominal shear strength of any one of the individual wall piers shall not be assumed to exceed $10A_{cp}\sqrt{f'_c}$, where A_{cp} represents the cross-sectional area of the pier considered. **1921.6.5.6**	No similar requirement.
	Nominal shear strength of horizontal wall segments shall not be assumed to exceed $10A_{cp}\sqrt{f'_c}$, where A_{cp} represents the cross-sectional area of a horizontal wall segment. **1921.6.5.7**	This limitation also exists in Zones 0, 1 & 2, except A_{cp} is replaced by *hd*. **1911.10.3**

A similar analysis is required for Seismic Zone 2 except that the nominal moment strength (M_n) of the section is used instead of the probable moment strength. For Zone 2, an alternative procedure is provided where the earthquake effects are doubled in lieu of using the nominal moment strength.

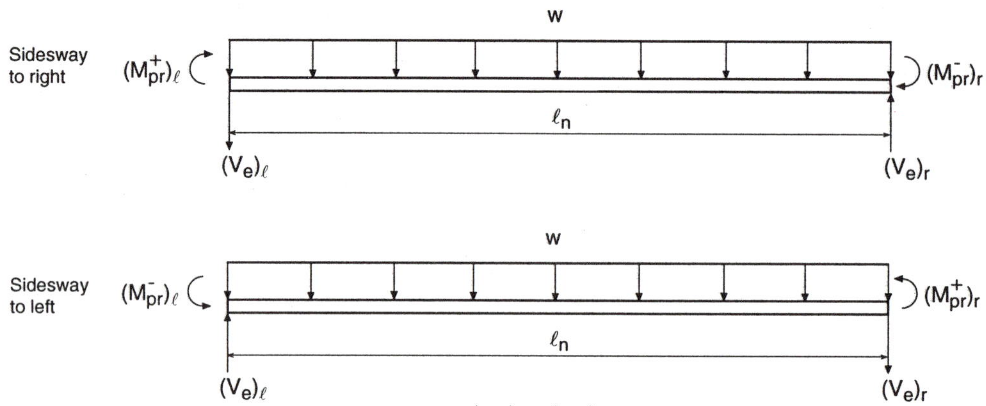

$$\text{Maximum } V_e = \frac{(M_{pr}^{\pm})_\ell + (M_{pr}^{\mp})_r}{\ell_n} + \frac{w\ell_n}{2}$$

For Zone 2, use nominal moment strength (M_n).

Figure 3-13 Design Shear Forces for Beams (Seismic Zones 2, 3 and 4)

In Seismic Zones 3 and 4, if the shear force from seismic loading is one-half or more of the maximum required shear strength, the contribution of the shear capacity from the concrete is taken as zero if the factored axial compressive force, including earthquake effects, is less than $A_g f_c' / 20$. The purpose of this requirement is not to imply that the concrete does not contribute to the shear carrying capacity, but rather to provide adequate shear reinforcement to ensure a flexural failure.

3.9.2 Members Subjected to Bending and Axial Loads

These requirements are the same as for flexural members. See Fig. 3-14 for the computation of the shear force (V_e).

3.9.3 Walls

The requirements for shear in walls are similar for all Seismic Zones (see Table 3-9). In Seismic Zones 0, 1, and 2, two alternate methods are given to compute the shear strength contributed by the concrete (V_c).

3.10 Two-Way Slabs Without Beams

Two-way slabs without beams are acceptable as LFR systems in regions of low and moderate seismicity (Zones 0, 1, and 2). Special seismic detailing provisions are included in 1921.8.6 for two-way slab systems without beams in regions of moderate seismicity (Zone 2). Note that this system is not permitted in Zones 3 and 4.

Table 3-10 summarizes the detailing requirements for two-way slab systems. Detailing requirements for two-way slabs without beams are shown in Figs. 3-15 and 3-16.

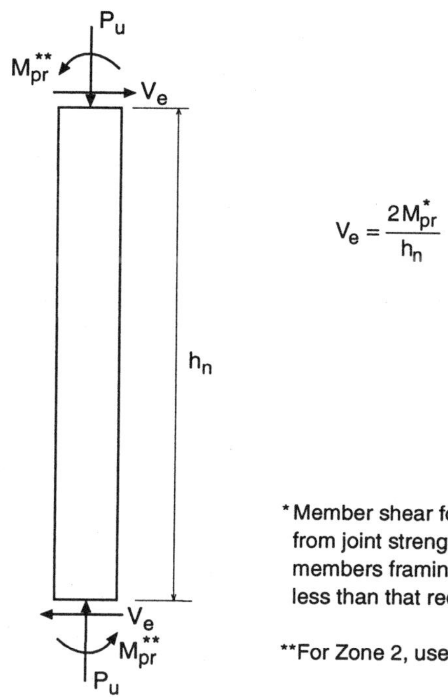

*Member shear forces need not exceed those determined from joint strengths based on M_{pr} of the transverse members framing into the joint. V_e shall never be taken less than that required by analysis of the structure.

**For Zone 2, use nominal moment strength (M_n).

Figure 3-14 Design Shear Forces for Columns (Seismic Zones 2, 3, and 4)

3.11 FRAME MEMBERS NOT PROPORTIONED TO RESIST FORCES INDUCED BY EARTHQUAKE MOTIONS

Special requirements for frame members that are assumed not to contribute to the lateral resistance are given in 1921.7 for buildings in Seismic Zones 2, 3, and 4. Specifically, these members are detailed depending on the magnitude of the moments that are induced when they are subjected to the lateral displacements due to the lateral forces. This requirement is intended to enable the gravity load system to maintain its vertical load carrying capacity under the design earthquake displacements. Although it is considered unnecessary that these elements of the gravity load system be designed for moments and shears related to the lateral forces, it has been recognized that they have to be designed to be adequate for the vertical loads and also the induced moments resulting from the design story drift.

The 1994 Northridge earthquake caused the collapse or partial collapse of at least two parking structures that could be attributed primarily to the failure of interior columns designed to carry gravity loads only. As a result, the requirements for frame members not proportioned to resist forces induced by earthquake motions (the so-called deformation compatibility requirements) have been extensively rewritten. The following summarizes these requirements:

(1) Compute moments and shears (E) in all elements that are not part of the LFR system due to the maximum inelastic response displacement (Δ_M) of the LFR system. The displacement (Δ_M) of the LFR system is determined for the code-prescribed forces per 1630.9.2.

*Table 3-10 Two-Way Slabs Without Beams**

Seismic Zones 0, 1 & 2
Zone 2 — All reinforcement provided to resist M_s, the portion of slab moment balanced by the support moment, shall be placed within the column strip. **1921.8.6.1** Zones 0 & 1 — In these zones, the middle strip is allowed to carry a portion of the unbalanced moment.
Zone 2 — The fraction, defined by Eq. (13-1), of the moment M_s shall be resisted by reinforcement placed within the effective width specified in 1913.5.3.2. **1921.8.6.2** Zones 0 & 1 — Similar requirement.
Zone 2 — Not less than one-half of the reinforcement in the column strip at the support shall be placed within the effective slab width specified in 1913.5.3.2. **1921.8.6.3** Zones 0 & 1 — No similar requirement.
Zone 2 — Not less than one-fourth of the top reinforcement at the support in the column strip shall be continuous throughout the span. **1921.8.6.4** Zones 0 & 1 — No similar requirement.
Zones 0, 1 & 2 — All bottom bars in the column strip shall be continuous or spliced with Class A splices. At least two of the column strip bottom bars shall pass within the column core and shall be anchored at exterior supports.† **1913.3.8.5** Zone 2 — Not less than one-half of all bottom reinforcement at midspan shall be continuous and shall develop its yield strength at the face of the support as defined in 1913.6.2.5. **1921.8.6.6** Zones 0 & 1 — No similar requirement.
Zone 2 — At discontinuous edges of the slab, all top and bottom reinforcement at the support shall be developed at the face of support as defined in 1913.6.2.5. **1921.8.6.7** Zones 0 & 1 — Positive moment reinforcement perpendicular to a discontinuous edge shall extend to the edge of the slab and have embedment at least 6 in. in spandrel beams, columns, or walls. Negative moment reinforcement perpendicular to a discontinuous edge must be anchored and developed at the face of the support according to provisions in 1912. **1913.3.3, 1913.3.4**

**Not permitted as part of the lateral force resisting system in Zones 3 & 4.*
†There appears to be a discrepancy between the provisions of 1921.8.6.5 and 1913.3.8.5 for bottom bars in the column strip. The requirements of 1913.3.8.5 are used here.

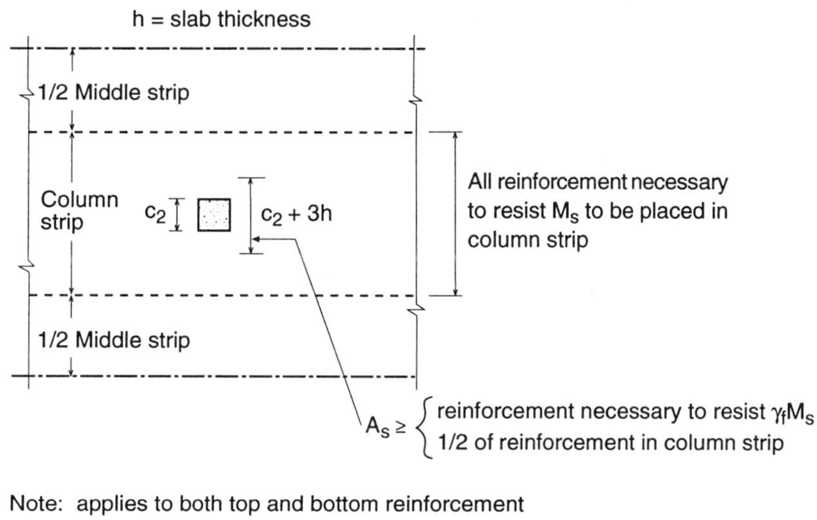

Figure 3-15 Location of Reinforcement in Slabs without Beams

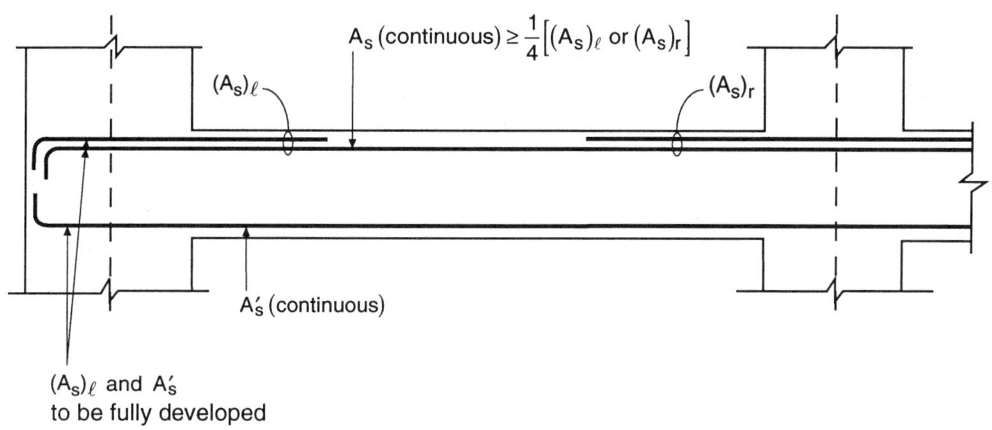

(a) Column Strip

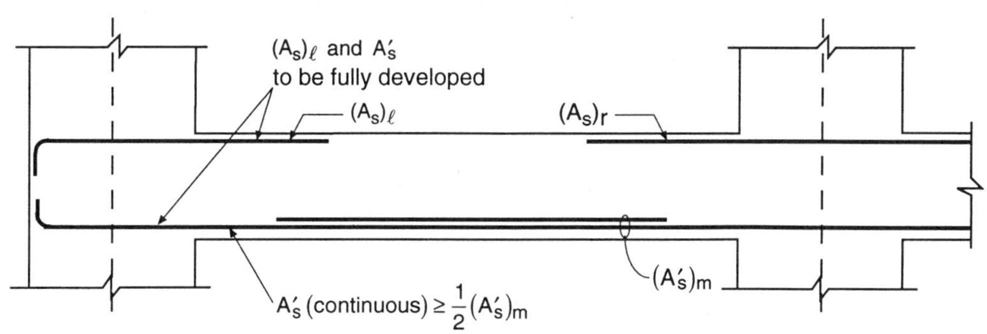

(b) Middle Strip

Figure 3-16 Details of Reinforcement in Slabs without Beams

(2) Determine the factored moment (M_u) and the factored shear (V_u) in each of the elements that are not part of the LFR system from the more critical of the following load combinations:

$$U = 1.4D + 1.4L + E$$

$$U = 0.9D + E$$

Note that the E-values (moments and shears) in the above expressions are determined in step 1 above.

(3) If $M_u \le \phi M_n$ and $V_u \le \phi V_n$ for an element that is not part of the LFR system, and if such an element is a beam $\left(P_u \le A_g f_c' / 10\right)$, it must satisfy the longitudinal reinforcement requirements in 1921.3.2.1; in addition, stirrups spaced at no more than $d/2$ must be provided throughout the length of the member. If such an element is a column $\left(P_u > A_g f_c' / 10\right)$ and $P_u \le 0.3P_o$ (where P_o is the nominal axial load strength at zero eccentricity), it must conform to 1921.4.3, 1921.4.4.1(3), and 1921.4.4.3. In addition, ties at a maximum spacing of s_o must be provided throughout the height of the column, where s_o must not exceed: (1) six times the smallest longitudinal bar diameter; (2) 16 tie bar diameters; (3) one-half the least cross-sectional dimension; and (4) 6 in. If the factored gravity axial force $P_u > 0.3P_o$, the requirements of 1921.4.4 and 1921.4.5 must be satisfied.

(4) If M_u or V_u determined in step 2 for an element that is not part of the LFR system exceeds ϕM_n or ϕV_n, or if induced moments and shears due to the lateral displacements are not calculated, then the structural materials must satisfy 1921.2.4 and 1921.2.5, and splices of reinforcement must satisfy 1921.2.6. If such an element is a beam $\left(P_u \le A_g f_c' / 10\right)$, it must conform to 1921.3.2.1 and 1921.3.4; in addition, stirrups spaced at no more than $d/2$ must be provided throughout the length of the member. If such an element is a column $\left(P_u > A_g f_c' / 10\right)$, it must be provided with full ductile detailing in conformance with 1921.4.4, 1921.4.5, and 1921.5.2.1.

3.12 REFERENCES

3.1 *Building Code Requirements for Structural Concrete (ACI 318-95) and Commentary (ACI 318R-95)*, American Concrete Institute, Farmington Hills, MI, 1995.

3.2 *Uniform Building Code*, International Conference of Building Officials, Whittier, CA, 1997.

4 BEAM-COLUMN FRAME

4.1 INTRODUCTION

A typical plan and elevation of a 3-story reinforced concrete storage warehouse building (Group S occupancy) is shown in Fig. 4-1. The computation of the seismic design forces according to the 1997 edition of the Uniform Building Code (UBC) is illustrated below. Typical beams and columns are designed and detailed for the combined effects of gravity, seismic, and wind forces.

A moment-resisting (beam-column) frame system is a structural system with an essentially complete frame that provides support for the gravity loads. Resistance to lateral loads is provided by the flexural action of the beams and columns. It is important to note that in Seismic Zones 3 and 4, frames that are part of the lateral force resisting system must be designed and detailed as special moment-resisting frames (1633.2.7). Intermediate and ordinary moment-resisting frames are required in Seismic Zones 2 and 1, respectively.

It is assumed mainly for simplicity that the columns have constant cross-sections throughout the height of the building, and that the bases of the lowest story segments are fixed. The beams and slabs are also assumed to have the same dimensions at all floor levels. Although the member dimensions in this example are within the practical range, the structure itself is a hypothetical one, and has been chosen mainly for illustrative purposes.

4.2 DESIGN DATA

- Material Properties
  Concrete: f'_c = 4,000 psi, w_c = 145 pcf
  Reinforcement: f_y = 60,000 psi

- Service Loads

  Live loads: roof = 20 psf
  floor = 125 psf (light storage; Table 16-A)

  Superimposed dead loads: roof = 10 psf
  floors = 15 psf
  Non-load bearing architectural precast panel: 5 in., w_c = 120 pcf
  Windows = 8 psf

- Seismic Design Data

  Seismic Zone 3 (Z = 0.30; see Table 16-I)
  Soil Profile Type S_D (Table 16-J)

For standard occupancy, Seismic Importance Factor (I) = 1.0 (Table 16-K)

For the special moment-resisting frame (SMRF) system, R = 8.5 (Table 16-N)

- Wind Design Data

Basic wind speed = 80 mph

Exposure C

For standard occupancy, Wind Importance Factor (I_w) = 1.0 (Table 16-K)

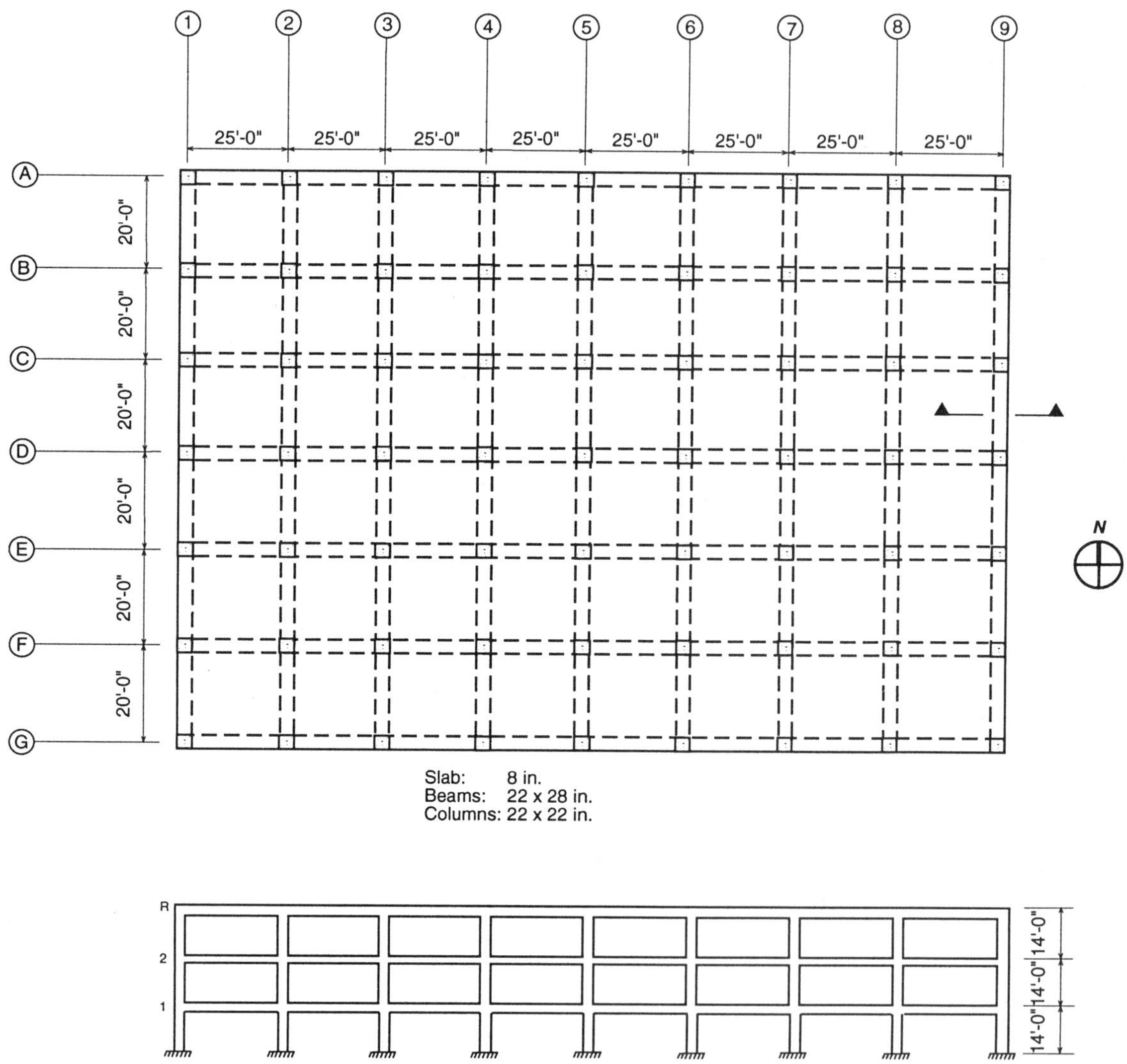

Figure 4-1 Typical Floor Plan and Elevation of Example Building

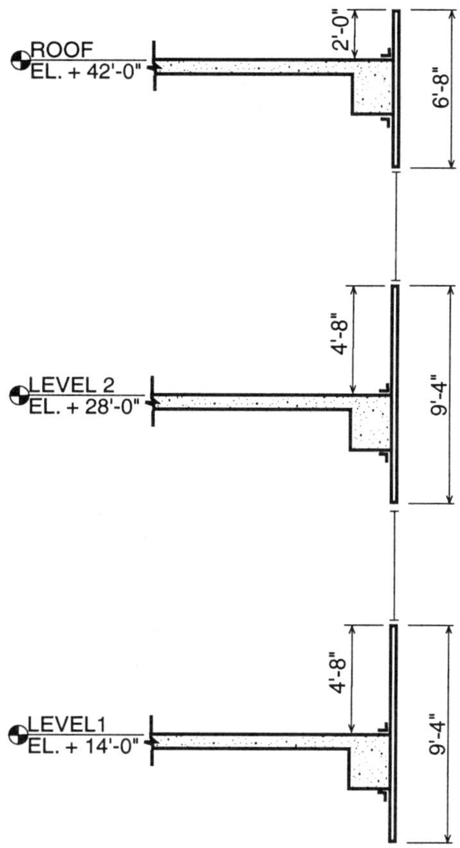

Figure 4-1 Typical Floor Plan and Elevation of Example Building (cont'd)

4.3 Seismic Load Analysis

4.3.1 Seismic Forces

The seismic base shear (V) is determined below according to the static force procedure (1630.2). Note that this procedure is applicable for this regular building which is less than 240 ft in height (1629.8.3). The base shear in is then distributed vertically to each floor level according to 1630.5.

Seismic Dead Load (*W*).

The dead loads for the roof level, the second floor level, and the first floor level are shown in Table 4-1. Note that for storage and warehouse occupancies, a minimum of 25% of the floor live load must be included in the seismic dead load (1630.1.1). The total seismic dead load of the building (W) is equal to $4{,}065 + (2 \times 5{,}250) = 14{,}565$ kips.

Table 4-1 Dead Loads at Various Floor Levels (kips)

	Beams	Slab*	25% Live Load	Vertical Elements**	Total
Roof	1,009	2,623	—	433	4,065
Floor Level 2	1,009	2,743	762	736	5,250
Floor Level 1	1,009	2,743	762	736	5,250

**includes the weight of the slab and the superimposed dead load*

***includes the weight of the columns, precast panels, and windows*

Approximate Period (*T*).

Method A in 1630.2.2 is used to compute the period (T). For the moment-resisting frame system:

Building height (h_n) = 42 ft

Numerical coefficient $C_t = 0.030$

Period $T = C_t\,(h_n)^{3/4} = 0.030 \times (42)^{3/4} = 0.495$ sec

Base Shear (*V*).

The total design base shear (V) is determined from Eq. (30-4) using $T = 0.495$ sec:

$$V = \frac{C_v IW}{RT}$$

where $C_v = 0.54$ for Seismic Zone 3 and soil profile type S_D (see Table 16-R).

Therefore,

$$V = \frac{0.54 \times 1.0 \times 14{,}565}{8.5 \times 0.495} = 1{,}869 \text{ kips}$$

The total design base shear need not exceed the value from Eq. (30-5):

$$V = \frac{2.5C_a IW}{R} = \frac{2.5 \times 0.36 \times 1.0 \times 14{,}565}{8.5} = 1{,}542 \text{ kips}$$

where $C_a = 0.36$ for Seismic Zone 3 and soil profile type S_D (see Table 16-Q).

Also, V shall not be less than the following value from Eq. (30-6):

$$V = 0.11C_a IW = 0.11 \times 0.36 \times 1.0 \times 14{,}565 = 577 \text{ kips}$$

Thus, use V = 1,542 kips. Note that this base shear is the same in both the N-S and E-W directions.

Lateral Forces (F_x).

The total base shear (V) is distributed over the height of the building in conformance with Eqs. (30-13), (30-14), and (30-15). The lateral force (F_x) at each floor level and the story shears (V_x) are contained in Table 4-2.

Table 4-2 Seismic Forces and Story Shears

Level	Story weight, w_x (kips)	Height, h_x (ft)	$w_x h_x$	Lateral Force, F_x* (kips)	Story Shear, V_x (kips)
3	4,065	42	170,730	673	673
2	5,250	28	147,000	579	1,252
1	5,250	14	73,500	290	1,542
Σ	14,565		391,230		

$$* F_x = \frac{(V - F_t)w_x h_x}{\sum_{i=1}^{n} w_i h_i} \quad \textit{(see 1630.5)}$$

Note that at the top of the building, the concentrated force F_t is equal to zero since the period of the building is less than 0.7 seconds (1630.5). Thus, the total force at the roof level is:

$$\text{Total force at roof level} = F_n + F_t = \frac{(V - F_t)w_n h_n}{\sum_{i=1}^{n} w_i h_i} + F_t = \frac{(1{,}542 - 0)\times 170{,}730}{391{,}230} + 0 = 673 \text{ kips}$$

4.3.2 Analysis Results

A three-dimensional analysis of the building was performed in the N-S and E-W directions for the seismic forces contained in Table 4-2 using *PCA-Frame* (Ref. 4.1). In the model, rigid diaphragms were assigned at each floor level, and rigid-end offsets were defined at the ends of each horizontal member so that results were automatically obtained at the faces of each support. For the beams, the portion of the adjoining slab defined in 1913.2.4 was included in the analysis as being part of the beam. In order to satisfy the modeling requirements of 1630.1.2, the stiffness properties of the members were input assuming cracked sections. In lieu of a more accurate analysis, the provisions of 1910.11.1 were used, i.e., the moment of inertia for the beams was taken as $0.35I_g$ where I_g is the gross moment of inertia of the section, and for the columns, it was taken as $0.70I_g$. Additionally, the modulus of elasticity was computed from 1908.5.1, and the gross area of the members was used in the analysis per 1910.11.1. In accordance with 1630.6, the mass at each level was assumed to be displaced from the calculated center of mass a distance equal to 5 percent of the building dimension at that level perpendicular to the direction of the force under consideration.

A summary of the results of the analysis for the interior frame along column line 2 for seismic forces in the N-S direction can be found in Fig. 4-2.

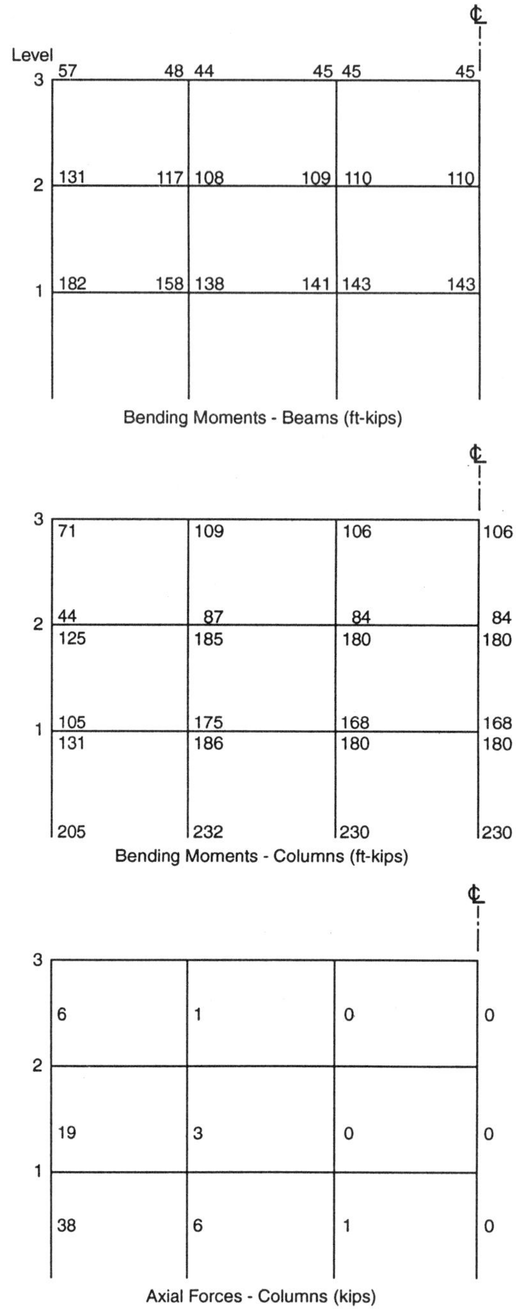

Figure 4-2 Analysis Results Along Column Line 2 for Seismic Forces in the N-S Direction

4.3.3 Story Drift Limitation and P-Δ Effects

Table 4-3 contains the displacements (Δ_S) obtained from the 3-D static, elastic analyses using the design seismic forces in the N-S direction, including accidental torsional effects. The table also contains the maximum inelastic response displacements (Δ_M) which are computed from Eq. (30-17):

$$\Delta_M = 0.7 R \Delta_S$$

where R is equal to 8.5. The interstory drifts computed from the Δ_S and Δ_M are also contained in the table. Note that the displacements in the E-W direction due to the seismic forces are negligible.

According to 1630.10.2, the calculated story drifts using Δ_M shall not exceed 0.025 times the story height for structures having a fundamental period less than 0.7 seconds. For the 14 ft story height, the maximum drift is equal to $0.025 \times 14 \times 12 = 4.2$ in. It is clear from Table 4-3 that for all stories, the lateral drifts obtained from the prescribed lateral forces are less than the limiting value.

According to 1630.1.3, P-Δ effects need not be considered for Seismic Zones 3 and 4 when the calculated story drift using Δ_S is less than or equal to $0.02h_s/R$, which is $0.02 \times 14 \times 12/8.5 = 0.40$ in. It can be seen from Table 4-3 that the interstory drifts are less than 0.40 in. so that P-Δ effects need not be considered.

Table 4-3 Lateral Displacements and Interstory Drifts due to Seismic Forces in the N-S Direction (in.)

Story	Δ_S	Drift	Δ_M	Drift
3	0.63	0.15	3.75	0.89
2	0.48	0.25	2.86	1.49
1	0.23	0.23	1.37	1.37

4.4 Wind Load Analysis

Table 4-4 contains the wind loads in the N-S direction computed according to 1609 with the basic wind speed equal to 80 mph, exposure C, and a wind importance factor (I_w) equal to 1.0.

Table 4-4 Design Wind Forces in N-S Direction (for entire structure)

Level	Height h_x (ft)	Tributary Height (ft)	Windward C_e	Windward Pressure* (psf)	Windward Force** (kips)	Leeward C_e	Leeward Pressure* (psf)	Leeward Force** (kips)	Total Force (kips)
3	42	7	1.33	17.4	24.6	1.33	10.9	15.4	40.0
2	28	14	1.21	15.9	45.0	1.33	10.9	30.7	75.7
1	14	14	1.06	13.9	39.3	1.33	10.9	30.7	70.0

*Windward pressure = $C_e C_q q_s I_w = C_e \times 0.8 \times 16.4 \times 1.0 = 13.12C_e$ psf

Leeward pressure = $C_e C_q q_s I_w = 1.33 \times 0.5 \times 16.4 \times 1.0 = 10.9$ psf

**Force = pressure × tributary height × 201.83/1,000

A summary of the results of the three-dimensional analysis for the interior frame along column line 2 for wind forces in the N-S direction is contained in Fig. 4-3.

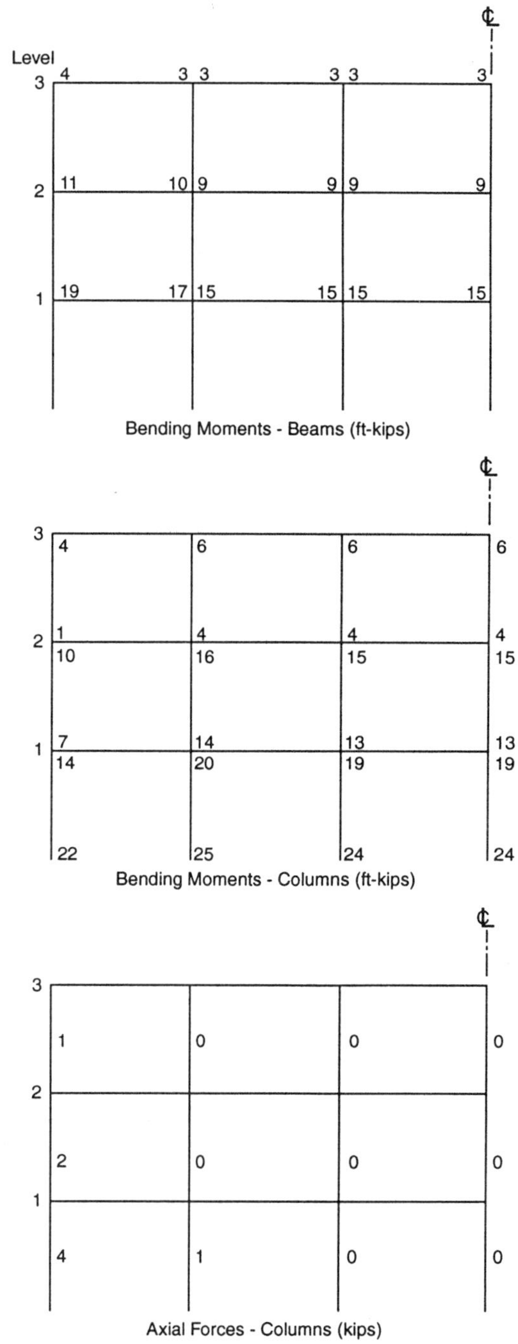

Figure 4-3 Analysis Results Along Column Line 2 for Wind Forces in the N-S Direction

4.5 GRAVITY LOAD ANALYSIS

4.5.1 Slab and Beams

The Direct Design Method (1913.6) is used to compute the gravity load moments in the slab and beams on a typical floor along column line 2. Note that this method is approximate and can only be used if certain geometric and loading criteria are satisfied (see 1913.6.1). In this case, all of the criteria are satisfied, including 1913.6.1.6:

$$5.0 \geq \frac{\alpha_1 \ell_2^2}{\alpha_2 \ell_1^2} \geq 0.2$$

where $\alpha_1, \alpha_2 = \alpha$ in direction of ℓ_1, ℓ_2

$\alpha = I_b / I_s$

I_b = moment of inertia of beam, including portion of adjoining slab defined in 1913.2.4

I_s = moment of inertia of slab

ℓ_1 = length of span in direction moments are being determined, measured center-to-center of supports

ℓ_2 = length of span transverse to ℓ_1, measured center-to-center of supports

For the N-S interior beams (see Fig. 4-4a):

$$I_b = \frac{1}{12}(62)(8)^3 + (62 \times 8)(24 - 17.42)^2 + \frac{1}{12}(22)(20)^3 + (22 \times 20)(17.42 - 10)^2 = 63{,}012 \text{ in.}^4$$

$$I_s = \frac{1}{12}(25 \times 12)(8)^3 = 12{,}800 \text{ in.}^4$$

$$\alpha = \frac{I_b}{I_s} = 4.9$$

Similarly, for the E-W interior beams:

$$I_b = 63{,}012 \text{ in.}^4$$

$$I_s = \frac{1}{12}(20 \times 12)(8)^3 = 10{,}240 \text{ in.}^4$$

$$\alpha = \frac{I_b}{I_s} = 6.2$$

For the E-W edge beams (see Fig. 4-4b):

$$I_b = \frac{1}{12}(42)(8)^3 + (42 \times 8)(24 - 16.06)^2 + \frac{1}{12}(22)(20)^3 + (22 \times 20)(16.06 - 10)^2 = 53{,}800 \text{ in.}^4$$

$$I_s = \frac{1}{12}(10.92 \times 12)(8)^3 = 5{,}591 \text{ in.}^4$$

$$\alpha = \frac{I_b}{I_s} = 9.6$$

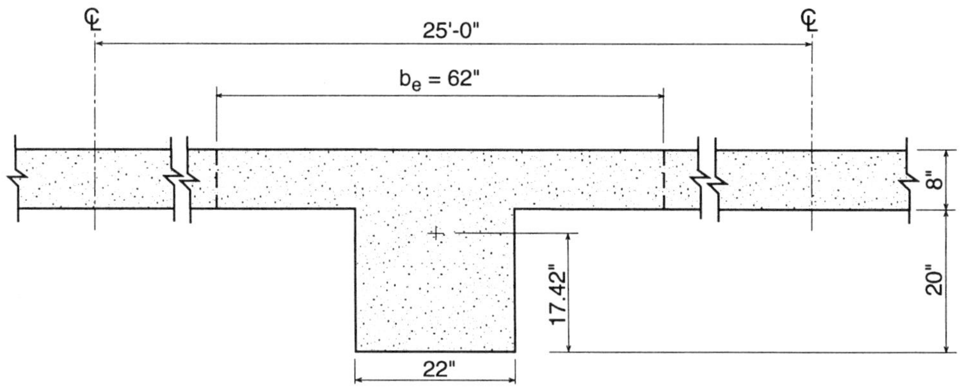

(a) Interior Beam

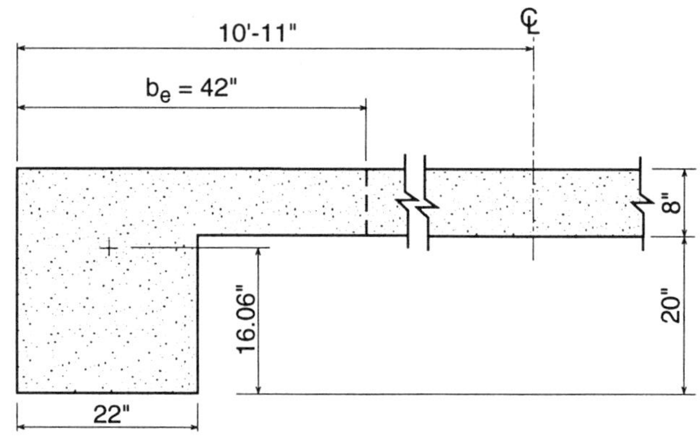

(b) Edge Beam

Figure 4-4 Section Properties of Interior and Edge Beams for Stiffness Computations

Thus, for an interior panel with the direction of analysis in the N-S direction:

$$\frac{\alpha_1 \ell_2^2}{\alpha_2 \ell_1^2} = \frac{4.9 \times 25^2}{6.2 \times 20^2} = 1.2, \ 5.0 > 1.2 > 0.2 \quad \text{O.K.}$$

For an edge panel:

$$\frac{\alpha_1 \ell_2^2}{\alpha_2 \ell_1^2} = \frac{4.9 \times 25^2}{9.6 \times 20^2} = 0.8, \ 5.0 > 0.8 > 0.2 \quad \text{O.K.}$$

The bending moments are determined using the dead and live loads per square foot of panel:

$$\text{Dead load} = \frac{8}{12}(145) + 15 + \frac{22 \times 20}{144}(145) \times \frac{1}{25} = 129 \text{ psf}$$

$$\text{Live load} = 125 \text{ psf}$$

Using Eq. (13-3), the total static service dead and live load moments for both an interior and end span in the N-S direction are computed as follows:

$$M_o = \frac{w\ell_2\ell_n^2}{8}$$

$$(M_o)_{DL} = \frac{0.129 \times 25 \times 18.17^2}{8} = 133.0 \text{ ft- kips}$$

$$(M_o)_{LL} = \frac{0.125 \times 25 \times 18.17^2}{8} = 128.9 \text{ ft- kips}$$

These moments must be divided into positive and negative moments, and then into column strip and middle strip moments according to the distribution factors given in 1913.6.3, 1913.6.4, and 1913.6.6. The column strip moments are further divided into beam and slab moments. The beams are designed for 85% of the column strip moments since $\alpha_1\ell_2/\ell_1$ = (4.9 × 25/20) = 6.1 > 1.0 (1913.6.5.1). The distribution factors for a two-way beam-supported slab are contained in Table 4-5.

It is important to note that the distribution factors for the exterior negative moments in the column strip are based on β_t greater than or equal to 2.5, where β_t is the ratio of the torsional stiffness of the edge beam to the flexural stiffness of the slab (see 1913.0). In this case, the torsional constant of the edge beam (C) is the larger value from the two cases shown in Fig. 4-5 (see 1913.0 for the definition of C):

Table 4-5 Design Moments at Face of Support for Two-Way Beam-Supported Slab

Span Ratio ℓ_2/ℓ_1	Moments		End Span Exterior Negative	End Span Positive	End Span Interior Negative	Interior Span Positive	Interior Span Interior Negative
	Total Moment		$0.16M_o$	$0.57M_o$	$0.70M_o$	$0.35M_o$	$0.65M_o$
0.5	Column Strip	Beam	$0.12M_o$	$0.43M_o$	$0.54M_o$	$0.27M_o$	$0.50M_o$
		Slab	$0.02M_o$	$0.08M_o$	$0.09M_o$	$0.05M_o$	$0.09M_o$
	Middle Strip		$0.02M_o$	$0.06M_o$	$0.07M_o$	$0.03M_o$	$0.06M_o$
1.0	Column Strip	Beam	$0.10M_o$	$0.37M_o$	$0.45M_o$	$0.22M_o$	$0.42M_o$
		Slab	$0.02M_o$	$0.06M_o$	$0.08M_o$	$0.04M_o$	$0.07M_o$
	Middle Strip		$0.04M_o$	$0.14M_o$	$0.17M_o$	$0.09M_o$	$0.16M_o$
2.0	Column Strip	Beam	$0.06M_o$	$0.22M_o$	$0.27M_o$	$0.14M_o$	$0.25M_o$
		Slab	$0.01M_o$	$0.04M_o$	$0.05M_o$	$0.02M_o$	$0.04M_o$
	Middle Strip		$0.09M_o$	$0.31M_o$	$0.38M_o$	$0.19M_o$	$0.36M_o$

Notes: (1) moments based on $\alpha_1\ell_2/\ell_1 \geq 1.0$ and $\beta_t \geq 2.5$.

(2) interpolate between values shown for different ℓ_2/ℓ_1 ratios.

(3) all negative moments are at face of support.

(4) concentrated loads applied directly to beams must be accounted for separately.

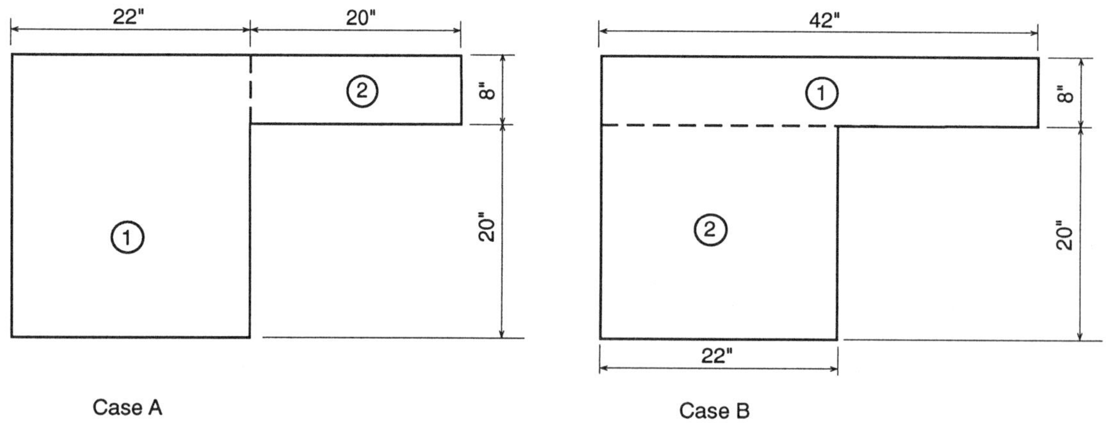

Figure 4-5 Component Rectangles to be Used in Computation of Torsional Constant (C)

- Case A:

$$C_1 = \left(1 - 0.63\frac{22}{28}\right)\left(\frac{22^3 \times 28}{3}\right) = 50{,}188 \text{ in.}^4$$

$$C_2 = \left(1 - 0.63\frac{8}{20}\right)\left(\frac{8^3 \times 20}{3}\right) = 2{,}553 \text{ in.}^4$$

$$C = C_1 + C_2 = 52{,}741 \text{ in.}^4 \quad \text{(governs)}$$

- Case B:

$$C_1 = \left(1 - 0.63\frac{8}{42}\right)\left(\frac{8^3 \times 42}{3}\right) = 6{,}308 \text{ in.}^4$$

$$C_2 = \left(1 - 0.63\frac{20}{22}\right)\left(\frac{20^3 \times 22}{3}\right) = 25{,}067 \text{ in.}^4$$

$$C = C_1 + C_2 = 31{,}375 \text{ in.}^4$$

In determining the moment of inertia of the slab (I_s), the width of the slab must equal the center-to-center length of the edge beam:

$$I_s = \frac{1}{12} \times (25 \times 12) \times (8)^3 = 12{,}800 \text{ in.}^4$$

Thus,

$$\beta_t = \frac{C}{2I_s} = \frac{52{,}741}{2 \times 12{,}800} = 2.1$$

Since β_t is less than 2.5, the distribution factors in Table 4-5 for the exterior negative moments in the column strip cannot be used. For the case when $\alpha_1 \ell_2 / \ell_1 > 1.0$, the following equations may be used to determine the percentage of total negative and positive bending moments to be resisted by the column strip (see Ref. 4.2):

Percentage of negative moment at exterior support to be resisted by the column strip

$$= 100 - 10\beta_t + 12\beta_t\left(1 - \frac{\ell_2}{\ell_1}\right) = 100 - (10 \times 2.1) + (12 \times 2.1)\left(1 - \frac{25}{20}\right) = 72.7\%$$

Percentage of negative moment at interior support to be resisted by the column strip

$$= 75 + 30\left(1 - \frac{\ell_2}{\ell_1}\right) = 75 + 30\left(1 - \frac{25}{20}\right) = 67.5\%$$

Percentage of positive moment to be resisted by the column strip

$$= 60 + 30\left(1.5 - \frac{\ell_2}{\ell_1}\right) = 60 + 30\left(1.5 - \frac{25}{20}\right) = 67.5\%$$

Thus, the total column strip moment at the exterior support in the end span is equal to $0.727 \times 0.16M_o = 0.12M_o$, with $0.85 \times 0.12M_o = 0.10M_o$ transferred to the beam and $0.02M_o$ transferred to the slab. The middle strip moment is $0.04M_o$ at this location. The moments at other locations along the span can be determined in a similar fashion. Table 4-6 contains the service gravity load moments for the slab and beams of an end span and an interior span on a typical floor.

Table 4-6 Service Dead and Live Load Moments (ft-kips) for Slab and Beams on a Typical Floor Along Column Line 2

			Moment	M_{DL}	M_{LL}
END SPAN					
Column Strip	Ext. Negative	Beam	$0.10M_o$	-13.3	-12.9
		Slab	$0.02M_o$	-2.7	-2.6
	Positive	Beam	$0.33M_o$	43.9	42.5
		Slab	$0.06M_o$	8.0	7.7
	Int. Negative	Beam	$0.40M_o$	-53.2	-51.6
		Slab	$0.07M_o$	-9.3	-9.0
Middle Strip	Ext. Negative		$0.04M_o$	-5.3	-5.2
	Positive		$0.18M_o$	23.9	23.2
	Int. Negative		$0.23M_o$	-30.6	-29.6
INTERIOR SPAN					
Column Strip	Positive	Beam	$0.20M_o$	26.6	25.8
		Slab	$0.04M_o$	5.3	5.2
	Int. Negative	Beam	$0.37M_o$	-49.2	-47.7
		Slab	$0.07M_o$	-9.3	-9.0
Middle Strip	Positive		$0.11M_o$	14.6	14.2
	Int. Negative		$0.21M_o$	-27.9	-27.1

4.5.2 Columns

The service axial loads for an interior and edge column along column line 2 are contained in Tables 4-7 and 4-8, respectively. Note that for storage loads exceeding 100 psf, the live loads may be reduced by 20% (1607.5). This 20% reduction is reflected in both tables for the first and second floor levels. The reduced live load for the columns supporting the roof level was computed in accordance with Eq. (7-1).

Table 4-7 Service Dead and Live Load Axial Forces for an Interior Column Along Column Line 2

Story	Dead Load* (psf)	Live Load (psf)	Supported Area (ft^2)	Reduced Live Load** (psf)	Live Load (kips)	Cumulative Dead Load† (kips)	Cumulative Live Load (kips)
3	107	20	500	14.4	7	79	7
2	112	125	1,000	100.0	50	160	57
1	112	125	1,500	100.0	50	241	107

*includes the weight of the slab and the superimposed dead load

**see 1607.5

†includes beam stem weight of 0.44 kips/ft and column weight of 0.49 kips/ft

Table 4-8 Service Dead and Live Load Axial Forces for an Edge Column Along Column Line 2

Story	Dead Load* (psf)	Live Load (psf)	Supported Area (ft^2)	Reduced Live Load** (psf)	Live Load (kips)	Cumulative Dead Load† (kips)	Cumulative Live Load (kips)
3	107	20	272.9	18.0	5	60	5
2	112	125	545.8	100.0	27	124	32
1	112	125	818.8	100.0	27	188	59

*includes the weight of the slab and the superimposed dead load

**see 1607.5

†includes beam stem weight of 0.44 kips/ft, column weight of 0.49 kips/ft, window weight of 0.2 kips/ft, and precast panel weight of 1.25 kips/ft

4.6 DESIGN FOR COMBINED LOAD EFFECTS

4.6.1 Load Combinations

The basic load combinations for the strength design method are given in 1612.2.1. According to the exceptions at the end of that section, the load combinations of 1909.2 are applicable for concrete where the load combinations do not include seismic forces. Also, the load combinations of 1612.2.1 which include seismic forces must be multiplied by 1.1 for concrete (see Sect. 2.2 for discussion on the 1.1 factor). Thus, the following load combinations are applicable to concrete members subjected to the combined effects of dead (D), live (L), wind (W), and earthquake (E) forces:

(1) $1.4D + 1.7L$

(2) $0.75(1.4D + 1.7L + 1.7W)$

(3) $0.9D + 1.3W$

(4) $1.1(1.2D + f_1 L + 1.0E) = 1.32D + 1.1 f_1 L + 1.1E$

(5) $1.1(0.9D + 1.0E) = 0.99D + 1.1E$

where f_1 = 1.0 for floors in places of public assembly, for live loads in excess of 100 psf, and for garage live loads
= 0.5 for other live loads

The earthquake force (E) is determined from Eq. (30-1):

$$E = \rho E_h + E_v$$

where E_h = earthquake force due to the base shear (V)

E_v = the load effect resulting from the vertical component of the earthquake motion

= $0.5C_aID = 0.5 \times 0.36 \times 1.0 \times D = 0.18D$ in this case

ρ = reliability/redundancy factor (see Eq. (30-3)):

$$1.0 \le \rho = 2 - \frac{20}{r_{max}\sqrt{A_B}} \le 1.25 \text{ for special moment-resisting frames}$$

r_{max} = maximum element-story shear ratio

A_B = ground floor area of the structure, ft^2

According to 1630.1.1, the maximum element-story shear ratio (r_{max}) is defined as the largest of the element-story shear ratios (r_i) which occurs in any of the story levels at or below the two-thirds height level of the building.

For moment-resisting frames, r_i is the maximum of the sum of shear forces in any two adjacent columns in a bay divided by the story shear (V_x). According to 1630.1.1, 70 percent of the shear force in a column that is common to two bays of the frame may be used in the shear force summation. The bay with the largest total shear force is to be used to determine r_i. Arbitrarily assuming that the center of mass is displaced to the west of its actual location, the frame along column line 1 will have the largest shear force. Table 4-9 contains the shear forces in the columns in the second and first story levels of frame 1. This table also contains the sum of the shear forces in adjacent columns and the element-story shear ratios (r_i). The maximum r_i occurs at the lowest level, so that r_{max} is equal to 0.030.

As was noted above, the reliability/redundancy factor (ρ) shall not be taken less than one. The value of r_{max} that corresponds to this lower limit can be determined from Eq. (30-3):

$$\rho = 2 - \frac{20}{r_{max}\sqrt{201.83 \times 121.83}} = 1.0 \rightarrow r_{max} = 0.128$$

Since r_{max} from Table 4-9 is less than this limiting value, ρ must be one. Thus, the seismic force (E) is:

$$E = E_h + 0.18D$$

Substituting E and f_1 = 1.0 (live load > 100 psf for first and second floors) into load combinations 4 and 5 above results in the following (see Sect 2.4.1 regarding the use of $0.18D$ in Eq. (5)):

(4) $1.32D + (1.1 \times 1.0)L + 1.1E = 1.32D + 1.1L + 1.1(E_h + 0.18D) = 1.52D + 1.1L + 1.1E_h$

(5) $0.99D + 1.1E = 0.99D + 1.1(E_h - 0.18D) = 0.79D + 1.1E_h$

Table 4-9 Element-Story Shear Ratios (r_i) for Seismic Forces in N-S Direction

Story	Column	Shear Force (kips)	Sum of Shear Forces in Adjacent Columns*	$r_i = \frac{\text{Max. sum of shear forces}}{\text{Story shear**}}$
	A1	17.4		
			36.0	
	B1	26.6		
			36.6	
	C1	25.7		
			36.1	
2	D1	25.8		36.6/1,252 = 0.029
			36.1	
	E1	25.7		
			36.6	
	F1	26.6		
			36.0	
	G1	17.4		
	A1	25.2		
			46.9	
	B1	31.0		
			42.9	
	C1	30.3		
			42.5	
1	D1	30.4		46.9/1,542 = 0.030
			42.5	
	E1	30.3		
			42.9	
	F1	31.0		
			46.9	
	G1	25.2		

**70% of the shear force in columns B1 through F1 is used in the summation (see 1630.1.1)*

***see Table 4-2 for story shear forces*

The values of E_h can be found in Fig. 4-2 for the beams and columns along column line 2. Also note that the special seismic load combinations of 1612.4 are not applicable in this example.

4.6.2 Beam and Slab Design

A summary of the slab and beam bending moments at the first floor level, based on the applicable load combinations, is contained in Table 4-10.

Design for Flexure – Slab.

The required reinforcement for the slab along column line 2 at the first floor level is summarized in Table 4-11. Since $\alpha_1 \ell_2 / \ell_1 > 1.0$, the beams carry the entire gravity load shear forces (see 1913.6.8). Reinforcement details for the slab are shown in Fig. 4-6.

Table 4-10 Summary of Slab and Beam Design Moments Along Column Line 2 at Floor Level 1

No.	Load Case	Location	End Span Moments (ft-kips) Column Strip Slab	End Span Column Strip Beam	End Span Middle Strip Slab	Interior Span Moments (ft-kips) Column Strip Slab	Interior Span Column Strip Beam	Interior Span Middle Strip Slab
	Dead (D)	Ext. Neg.	-2.7	-13.3	-5.3	—	—	—
		Positive	8.0	43.9	23.9	5.3	26.6	14.6
		Int. Neg.	-9.3	-53.2	-30.6	-9.3	-49.2	-27.9
	Live (L)	Ext. Neg.	-2.6	-12.9	-5.2	—	—	—
		Positive	7.7	42.5	23.2	5.2	25.8	14.2
		Int. Neg.	-9.0	-51.6	-29.6	-9.0	-47.7	-27.1
	Earthquake (E_h)	Ext. Neg.	—	± 182.0	—	—	—	—
		Positive	—	—	—	—	—	—
		Int. Neg.	—	± 158.0	—	—	± 138.0	—
	Wind (W)	Ext. Neg.	—	± 19.0	—	—	—	—
		Positive	—	—	—	—	—	—
		Int. Neg.	—	± 17.0	—	—	± 15.0	—
No.	**Load Combination**							
1	$1.4D + 1.7L$	Ext. Neg.	-8.2	-40.6	-16.3	—	—	—
		Positive	24.3	133.7	72.9	16.3	81.1	44.6
		Int. Neg.	-28.3	-162.2	-93.2	-28.3	-150.0	-85.1
2	$0.75(1.4D + 1.7L + 1.7W)$	Ext. Neg.	-6.2	-6.2	-12.2	—	—	—
				-54.6				
		Positive	18.2	100.3	54.7	12.2	60.8	33.4
		Int. Neg.	-21.2	-143.3	-69.9	-21.2	-93.4	-63.8
				-100.0			-131.6	
3	$0.9D + 1.3W$	Ext. Neg.	-2.4	12.7	-4.8	—	—	—
				-36.7				
		Positive	7.2	39.5	21.5	4.8	23.9	13.1
		Int. Neg.	-8.4	-70.0	-27.5	-8.4	-24.8	-25.1
				-25.8			-63.8	
4	$1.52D + 1.1L + 1.1E_h$	Ext. Neg.	-7.0	165.8	-13.8	—	—	—
				-234.6				
		Positive	20.6	113.5	61.8	13.8	68.8	37.8
		Int. Neg.	-24.0	-311.4	-79.1	-24.0	24.5	-72.2
				36.2			-279.1	
5	$0.79D + 1.1E_h$	Ext. Neg.	-2.1	189.7	-4.2	—	—	—
				-210.7				
		Positive	6.3	34.7	18.9	4.2	21.0	11.5
		Int. Neg.	-7.3	-215.8	-24.2	-7.3	112.9	-22.0
				131.8			-190.7	

Table 4-11 Required Slab Reinforcement at Floor Level 1

Location			M_u (ft-kips)	b (in.)	d (in.)	A_s (in.^{2})	$A_{s,min}$* (in.^{2})	Reinforcement**
End Span	Column Strip	Ext. Neg.	-8.2	120	6.75	0.27	1.73	9-No. 4
		Positive	24.3	120	6.75	0.81	1.73	9-No. 4
		Int. Neg.	-28.3	120	6.75	0.94	1.73	9-No. 4
	Middle Strip	Ext. Neg.	-16.3	180	6.75	0.54	2.59	13-No. 4
		Positive	72.9	180	6.75	2.44	2.59	13-No. 4
		Int. Neg.	-93.2	180	6.75	3.14	2.59	16-No. 4
Interior Span	Column Strip	Positive	16.3	120	6.75	0.54	1.73	9-No. 4
		Negative	-28.3	120	6.75	0.94	1.73	9-No. 4
	Middle Strip	Positive	44.6	180	6.75	1.48	2.59	13-No. 4
		Negative	-85.1	180	6.75	2.86	2.59	15-No. 4

*$A_{s,min}$ = 0.0018bh (1913.3.1)

**s_{max} = 2h = 16 in. (1913.3.2)

For b = 120 in., 120/16 = 7.5 spaces, say 9 bars. For b = 180 in., 180/16 = 11.3 spaces, say 13 bars.

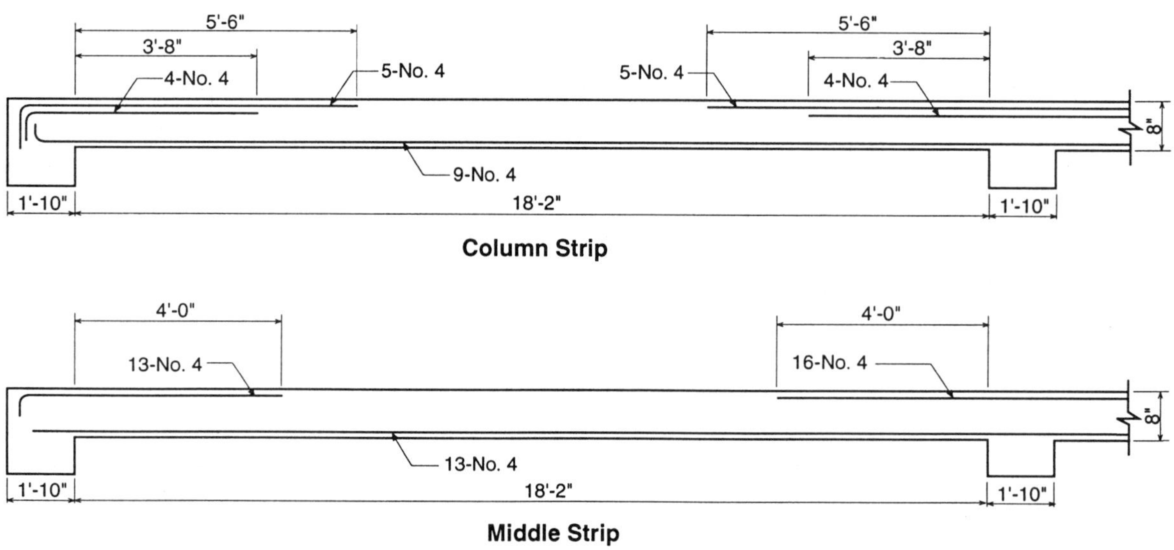

Figure 4-6 Reinforcement Details for Slab Along Column Line 2 on Floor Level 1

Design for Flexure – Beams.

According to 1921.2.1.4, in Seismic Zones 3 and 4, all reinforced concrete members that are part of the lateral load resisting system must satisfy the requirements of 1921.2 through 1921.7, in addition to the requirements of 1901 through 1917. In particular, flexural frame members must satisfy the requirements of 1921.3. A check of the conditions of 1921.3.1 is shown below:

- Factored axial compressive force on beams is negligible. O.K.
- Minimum clear span = 18.17 ft > $4d = 4 \times (28 - 2.5)/12 = 8.5$ ft O.K.
- $\dfrac{\text{Width}}{\text{Depth}} = \dfrac{22}{28} = 0.79 > 0.3$ O.K.

- Width = 22 in. > 10 in. O.K.
- Width = 22 in. < width of supporting column + (1.5 × depth of beam)

  < 22 + (1.5 × 28) = 64 in. O.K.

The required reinforcement for the beams on the first floor level is contained in Table 4-12. The provided areas of steel are within the limits specified in 1921.3.2.1. Also given in Table 4-12 are the flexural moment strengths (ϕM_n) at each section. According to 1921.3.2.2, the positive moment strength at a joint face must be at least equal to 50% of the negative moment strength provided at that joint. At the exterior negative location, this provision is satisfied since the positive moment strength of 265.0 ft-kips is greater than 265.0/2 = 132.5 ft-kips. The provision is also satisfied at the interior negative location since 265.0 ft-kips is greater than 328.0/2 = 164.0 ft-kips. Additionally, neither the negative nor the positive moment strength at any section along the length of the member shall be less than 25% of the maximum moment strength provided at the face of either joint (1921.3.2.2). In this case, 25% of the maximum moment strength is equal to 328.0/4 = 82.0 ft-kips. Providing at least 2-No. 7 bars (ϕM_n = 135.1 ft-kips) at any section will satisfy this requirement. However, to satisfy the minimum reinforcement requirement of 1921.3.2.1 (i.e., minimum A_s = 1.87 in.^{2}), a minimum of 4-No. 7 bars (A_s = 2.40 in.^{2}) must be provided at any section. This also automatically satisfies the requirement that 2 bars be continuous at both the top and bottom of any section (1921.3.2.1).

When reinforcing bars extend through a joint, the column dimension must be at least 20 times the diameter of the largest longitudinal bar for normal weight concrete (1921.5.1.4). In this case, the minimum required column dimension is 20 × 0.875 = 17.5 in. which is less than the 22 in. column widths that are provided.

Table 4-12 Required Reinforcement for Beams at Floor Level 1

Location		M_u (ft-kips)	A_s* (in.^{2})	Reinforcement*	ϕM_n (ft-kips)
End Span	Ext. Neg.	-234.6	2.11	4-No. 7	265.0
		189.7	1.87	4-No. 7	265.0
	Positive	133.7	1.87	4-No. 7	265.0
	Int. Neg.	-311.4	2.84	5-No. 7	328.0
		131.8	1.87	4-No. 7	265.0
Interior Span	Positive	81.1	1.87	4-No. 7	265.0
	Negative	-279.1	2.53	5-No. 7	328.0
		112.9	1.87	4-No. 7	265.0

**Max.* $A_s = 0.025 \times 22 \times 25.5 = 14.03$ *in.*^{2} *(1921.3.2.1)*

Min. $A_s = 3\sqrt{4{,}000} \times 22 \times 25.5/60{,}000 = 1.77$ *in.*^{2}

$= 200 \times 22 \times 25.5/60{,}000 = 1.87$ *in.*^{2} *(governs)*

Design for Shear – Beams.

The beams must be designed for shear forces in accordance with 1921.3.4. These shear forces are determined assuming that moments of opposite sign corresponding to the probable moment strength (M_{pr}) act at the joint faces and that the member is loaded with the tributary gravity load along its span.

It is important to note that sidesway to the right and to the left must both be considered when calculating the maximum design shear forces.

The probable moment strength (M_{pr}) is determined for a section using the stress in the tensile reinforcement equal to $1.25f_y$ and a strength reduction factor (ϕ) equal to 1.0. The following equation can be used to compute M_{pr} :

$$M_{pr} = A_s(1.25f_y)\left(d - \frac{a}{2}\right)$$

where $\quad a = \dfrac{A_s(1.25f_y)}{0.85f_c' b}$

For example, for sidesway to the right, the interior joint must be subjected to the negative moment M_{pr} which is determined as follows:

For 5-No. 7 top bars, $A_s = 5 \times 0.60 = 3.0$ in.^{2}

$$a = \frac{A_s(1.25f_y)}{0.85f_c' b} = \frac{3.0 \times 1.25 \times 60}{0.85 \times 4 \times 22} = 3.0 \text{ in.}$$

$$M_{pr} = A_s(1.25f_y)\left(d - \frac{a}{2}\right) = 3.0 \times 1.25 \times 60 \times \left(25.5 - \frac{3.0}{2}\right) = 5{,}400 \text{ in.-kips} = 450 \text{ ft-kips}$$

Similarly, for the exterior joint, the positive moment M_{pr} based on the 4-No. 7 bottom bars is equal to 364.5 ft-kips. The probable moment strengths for sidesway to the left can be obtained in a similar fashion.

The gravity load on a beam along column line 2 is:

$$w = \frac{8}{12}(145) + 15 + \frac{(20 \times 22) \times 145}{25 \times 144} + 125 = 254 \text{ psf}$$

For beams with $\alpha_1 \ell_2 / \ell_1 \geq 1.0$, the tributary area for shear is bounded by 45-deg lines drawn from the corners of the panel and the centerlines of the adjacent panels parallel to the long sides (1913.6.8.1). For the beams in the direction of analysis, the tributary area is triangular, as depicted in Fig. 4-7. At midspan, the magnitude of the gravity load is equal to $0.254 \times 20 = 5.08$ kips/ft.

Figure 4-8 shows the exterior beam and the shear forces due to the gravity loads. Also shown are the probable moment strengths (M_{pr}) at the joint faces for sidesway to the right and to the left and the corresponding shear forces due to these moments. Note that the maximum combined design shear forces are larger than those obtained from the structural analysis.

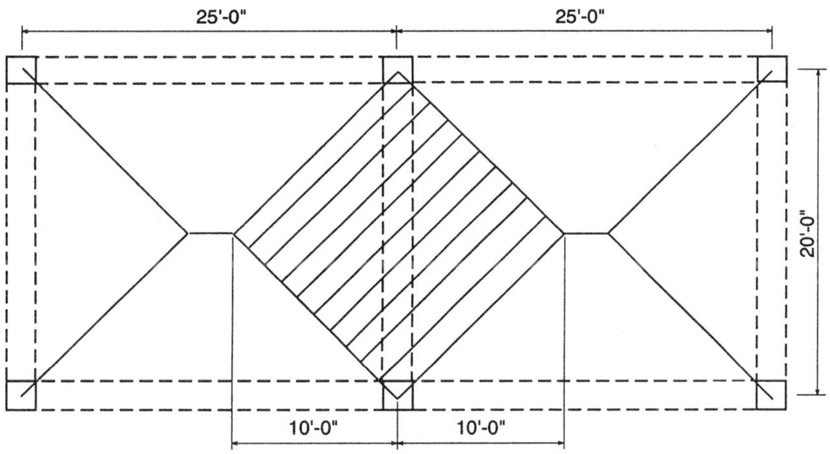

Figure 4-7 Tributary Area for Shear on an Interior Beam

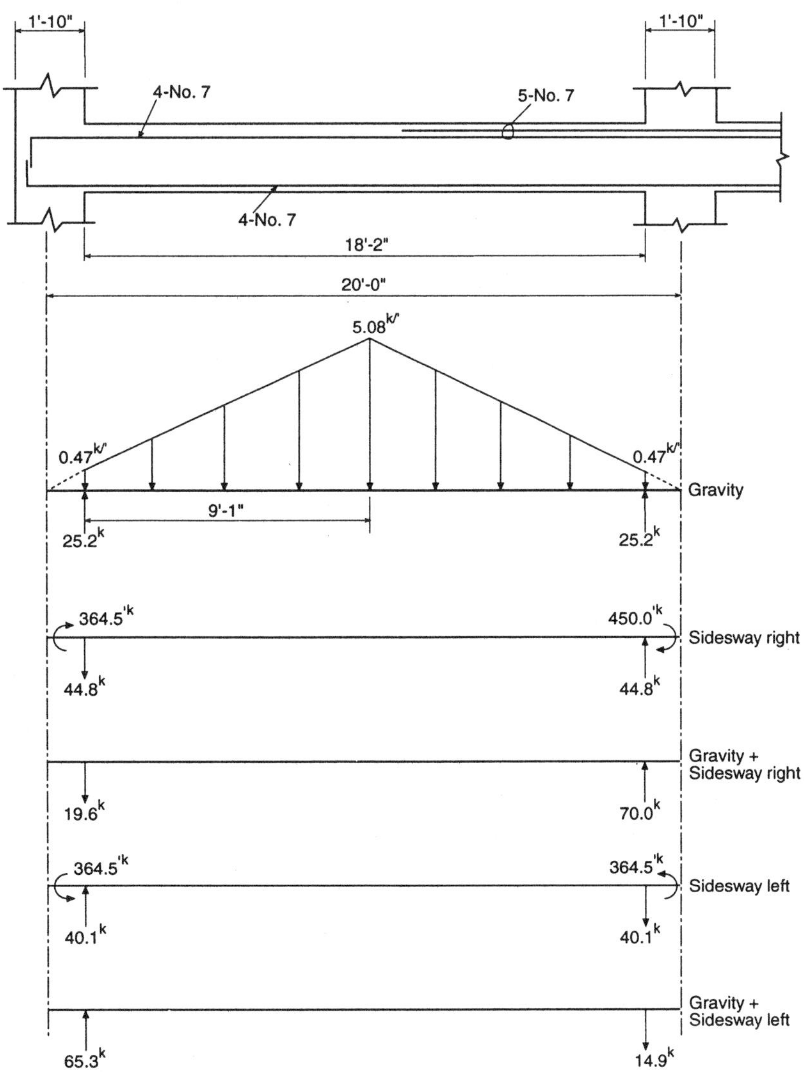

Figure 4-8 Design Shear Forces for Exterior Beam Along Column Line 2 on Floor Level 1

In general, the shear strength is assumed to be provided by both concrete (V_c) and reinforcing steel in the form of hoops or stirrups (V_s). However, the shear strength of the concrete (V_c) is to be taken as zero when the earthquake-induced shear force calculated in accordance with 1921.3.4.1 is greater than or equal to 50% of the total shear force and the factored axial compressive force is less than $A_g f'_c/20$ where A_g is the gross area of the beam (1921.3.4.2). In this particular example, the beam carries negligible axial forces, and the maximum earthquake-induced shear force, which is equal to 44.8 kips (see Fig. 4-8), is more than one-half the total design shear force which is equal to 0.5 × 70.0 = 35.0 kips. Thus, V_c must be zero. The maximum shear force V_s is (1911.1):

$$\phi V_s = V_u - \phi V_c$$

$$V_s = \frac{V_u}{\phi} - 0 = \frac{70.0}{0.85} = 82.4 \text{ kips}$$

where the strength reduction factor (ϕ) was taken as 0.85 since the shear force (V_u) corresponds to the nominal flexural strength of the member (see 1909.3.4.1).

Shear strength contributed by shear reinforcement shall not exceed $(V_s)_{max}$ (1911.5.6.8):

$$(V_s)_{max} = 8\sqrt{f'_c}\, b_w d = 8\sqrt{4{,}000} \times 22 \times 25.5/1{,}000 = 283.9 \text{ kips} > 82.4 \text{ kips} \quad \text{O.K.}$$

Also, V_s is less than $4\sqrt{f'_c}\, b_w d = 141.9$ kips.

Required spacing of No. 4 closed stirrups (hoops) for a factored shear force of 82.4 kips is (1911.5.6.2):

$$s = \frac{A_v f_y d}{V_s} = \frac{(3 \times 0.2) \times 60 \times 25.5}{82.4} = 11.1 \text{ in.}$$

Note that 3 legs are required in order to conform with the provisions of 1921.3.3.3 for lateral support of the longitudinal bars.

Maximum allowable hoop spacing (s_{max}) within a distance of $2h = 2 \times 28 = 56$ in. (1921.3.3.1) from the face of the support is the smallest of the following (1921.3.3.2):

$$s_{max} = \frac{d}{4} = \frac{25.5}{4} = 6.4 \text{ in. (governs)}$$

$$= 8 \times (\text{diameter of smallest longitudinal bar}) = 8 \times 0.875 = 7.0 \text{ in.}$$

$$= 24 \times (\text{diameter of hoop bar}) = 24 \times 0.50 = 12.0 \text{ in.}$$

$$= 12 \text{ in.}$$

Therefore, hoops must be spaced at 6 in. on center with the first one located at 2 in. from the face of the support. Ten hoops are to be placed at this spacing. Where hoops are no longer required, stirrups with seismic hooks at both ends may be used (1921.3.3.4). At a distance of 56 in. from the face of the interior support, V_u = 62.3 kips which requires No. 4 stirrups (2 legs) at s = 12.8 in. A 12 in. spacing, starting at 56 in. from the face of the support will be sufficient for the remaining portion of the beam. Note that this spacing is less than the maximum allowable spacing which is $d/2$ = 12.75 in. (1921.3.3.4).

Reinforcing Bar Cutoff Points – Beams.

The negative reinforcement at the interior support is 5-No. 7 bars. The location where one of the five bars can be terminated will be determined. Note that 4-No. 7 bars must be continuous throughout the length of the beam to satisfy the minimum reinforcement requirements of 1921.3.2.1.

The loading used to find the cutoff point of the 1-No. 7 bar at this location is 0.9 times the dead load in combination with the probable flexural strengths (M_{pr}) at the ends of the member; this combination will produce the longest bar lengths. The design flexural strength (ϕM_n) provided by 4-No. 7 bars is 265.0 ft-kips. Therefore, the 1 reinforcing bar can be terminated after the required moment strength (M_u) has been reduced to 265.0 ft-kips.

With $\phi = 1.0$ and $f_s = 75$ ksi, $M_{pr} = 450.0$ ft-kips at the interior support and 364.5 ft-kips at the exterior support. The load is equal to $0.9 \times (129 \times 20)/1{,}000 = 2.32$ kips/ft at midspan. The distance from the face of the interior support to where the moment equals 265.0 ft-kips is obtained by summing moments about section *a-a* in Fig. 4-9:

$$\frac{x}{2}\left[\frac{(2.32-0.21)x}{9.083}\right]\left(\frac{x}{3}\right)+\frac{0.21x^2}{2}-56.3x+450.0=265.0$$

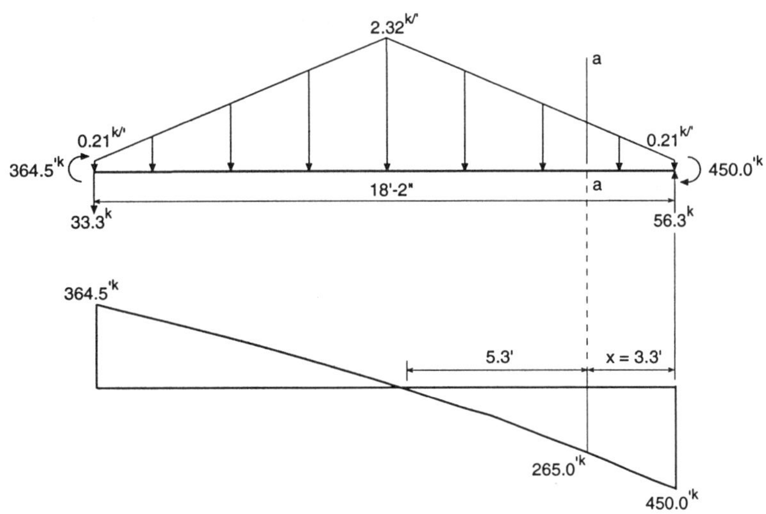

Figure 4-9 Moment Diagram for Cutoff Location of Negative Bars at Interior Support

Solving for x gives a distance of 3.3 ft. The 1-No. 7 bar must extend a distance $d = 25.5$ in. or $12d_b = 12 \times 0.875 = 10.5$ in. beyond the distance x (1912.10.3). Thus, from the face of the support, the total bar length must be at least equal to $3.3 + (25.5/12) = 5.4$ ft. Also, the bar must extend a full development length ℓ_d beyond the face of the support (1912.10.4). The development length can be determined from the provisions of 1912.2.3:

$$\frac{\ell_d}{d_b}=\frac{3}{40}\frac{f_y}{\sqrt{f_c'}}\frac{\alpha\beta\gamma\lambda}{\left(\frac{c+K_{tr}}{d_b}\right)}$$

where α = reinforcement location factor = 1.3 (top bar)

β = coating factor = 1.0 (uncoated reinforcement)

γ = reinforcement size factor = 1.0 (No. 7 bar)

λ = lightweight aggregate concrete factor = 1.0 (normal weight concrete)

$$c = \text{spacing or cover dimension} = \begin{cases} 1.5 + 0.5 + \dfrac{0.875}{2} = 2.4\text{ in.} \\ \dfrac{22 - 2(1.5 + 0.5) - 0.875}{2 \times 4} = 2.1\text{ in. (governs)} \end{cases}$$

K_{tr} = transverse reinforcement index = 0 (conservative)

$$\frac{c + K_{tr}}{d_b} = \frac{2.1 + 0}{0.875} = 2.4$$

$$\frac{\ell_d}{d_b} = \frac{3}{40} \times \frac{60{,}000}{\sqrt{4{,}000}} \times \frac{1.3 \times 1.0 \times 1.0 \times 1.0}{2.4} = 38.5$$

$$\ell_d = 38.5 \times 0.875 = 33.7\text{ in.} = 2.8\text{ ft} < 5.4\text{ ft}$$

Thus, the total required length of the 1-No. 7 bar must be at least 5.4 ft beyond the face of the support. It is important to note that flexural reinforcement shall not be terminated in a tension zone unless one or more of the conditions of 1912.10.5 are satisfied. In this case, the point of inflection is approximately 8.6 ft from the face of the right support which is greater than 5.4 ft. Therefore, the 1-No. 7 bar can not be terminated here unless one of the conditions of 1912.10.5.1 or 1912.10.5.2 are satisfied. In lieu of increasing the amount of shear reinforcement (1912.10.5.2), determine the location where the factored shear force (V_u) is equal to two-thirds of that permitted ($2\phi V_n/3$) and extend the flexural reinforcement to at least that location (1912.10.5.1). For No. 4 stirrups at 12 in. on center that are provided in this region of the beam:

$$\phi V_n = \phi(V_s + V_c) = 0.85 \times \left(\frac{0.4 \times 60 \times 25.5}{12} + 71.0\right) = 103.7\text{ kips}$$

$$\frac{2}{3}\phi V_n = 69.1\text{ kips} > V_u = 51.7\text{ kips at 5.4 ft from face of support}$$

Since $2\phi V_n/3 > V_u$, the cutoff point for the 1-No. 7 bar can be 5.4 ft from the face of the interior support.

Note that the reinforcing bars that terminate at exterior columns must be properly developed by providing 90-deg hooks embedded in the confined core of the column (1921.5.1.3). The development length of a hook (ℓ_{dh}) is the largest of the following (1921.5.4.1):

$$\ell_{dh} = 8d_b = 8 \times 0.875 = 7.0 \text{ in.}$$

$$= 6 \text{ in.}$$

$$= \frac{f_y d_b}{65\sqrt{f'_c}} = \frac{60{,}000 \times 0.875}{65\sqrt{4{,}000}} = 12.8 \text{ in. (governs)}$$

Therefore, the hooks must extend at least 13 in. into the column with a $12d_b$ = 10.5 in. extension (1907.1.2). Figure 4-10 shows the reinforcement details for the exterior beam.

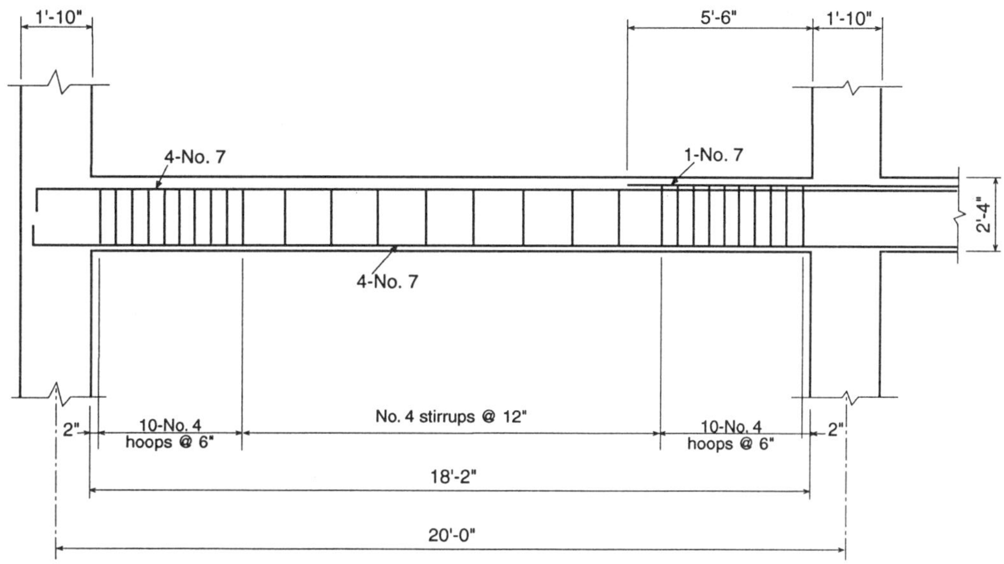

Figure 4-10 Reinforcement Details for Exterior Beam Along Column Line 2 on Floor Level 1

4.6.3 Column Design

This section outlines the design of an interior column and an edge column along column line 2 below the first floor level. A summary of the design forces is given in Tables 4-13 and 4-14 for column A2 and Tables 4-15 and 4-16 for column B2. Note that in Seismic Zones 2, 3 and 4, the orthogonal effects of the earthquake forces must be considered unless the axial load in the column due to seismic forces in either direction is less than 20% of the column axial load capacity (1633.1). Conservatively assuming a 1% percent reinforcement ratio for the 22 × 22 in. interior column, the axial load capacity ($\phi P_{n(max)}$) is equal to 1,075 kips (1910.3.5.1); 20% of this load is 215 kips which is greater than the axial load of 6 kips due to seismic forces in the N-S direction (see Fig. 4-2) and 5 kips due to seismic forces in the E-W direction. Similarly, for the edge column, 215 kips is greater than 38 kips and 5 kips which are the axial loads due to seismic forces in the N-S and E-W directions, respectively. Thus, orthogonal effects need not be considered.

The requirements of 1921.4 apply to members resisting earthquake forces and factored axial forces exceeding $A_g f'_c/10$. For a 22 × 22 in. column with f'_c = 4 ksi, $A_g f'_c/10$ = 194 kips which is less than the maximum factored axial load on a interior or edge column (see Tables 4-13 through 4-16). Thus, the following two criteria must be satisfied:

- Shortest cross-sectional dimension = 22 in. > 12 in. O.K.
- Ratio of shortest cross-sectional dimension to perpendicular dimension = 1.0 > 0.4 O.K.

Table 4-13 Summary of Factored Axial Loads and Bending Moments for Column B2 between the 1st and 2nd Floor Levels

No.	Load Case	Axial Load (kips)	Bending Moment (ft-kips)
	Dead (D)	160	0.0
	Live (L)	57	-18.1
	Earthquake (E_h)	3	185.0
	Wind (W)	0	16.0
No.	**Load Combination**		
1	$1.4D + 1.7L$	321	-30.8
2	$0.75(1.4D + 1.7L + 1.7W)$	241	-2.7
		241	-43.5
3	$0.9D + 1.3W$	144	20.8
		144	-20.8
4	$1.52D + 1.1L + 1.1E_h$	303	183.6
		309	-223.4
5	$0.79D + 1.1E_h$	123	203.5
		130	-203.5

Table 4-14 Summary of Factored Axial Loads and Bending Moments for Column B2 between the Ground and 1st Floor Levels

No.	Load Case	Axial Load (kips)	Bending Moment (ft-kips)
	Dead (D)	241	0.0
	Live (L)	107	-18.1
	Earthquake (E_h)	6	232.0
	Wind (W)	1	25.0
No.	**Load Combination**		
1	$1.4D + 1.7L$	519	-30.8
2	$0.75(1.4D + 1.7L + 1.7W)$	388	8.8
		391	-55.0
3	$0.9D + 1.3W$	216	32.5
		218	-32.5
4	$1.52D + 1.1L + 1.1E_h$	477	235.3
		491	-275.1
5	$0.79D + 1.1E_h$	184	255.2
		197	-255.2

Table 4-15 Summary of Factored Axial Loads and Bending Moments for Column A2 between the 1st and 2nd Floor Levels

No.	Load Case	Axial Load (kips)	Bending Moment (ft-kips)
	Dead (D)	124	-10.7
	Live (L)	32	-10.4
	Earthquake (E_h)	19	125.0
	Wind (W)	2	10.0
No.	**Load Combination**		
1	$1.4D + 1.7L$	228	-32.7
2	$0.75(1.4D + 1.7L + 1.7W)$	169	-11.7
		174	-37.2
3	$0.9D + 1.3W$	109	3.4
		114	-22.6
4	$1.52D + 1.1L + 1.1E_h$	203	109.8
		245	-165.2
5	$0.79D + 1.1E_h$	77	129.0
		119	-146.0

Table 4-16 Summary of Factored Axial Loads and Bending Moments for Column A2 between the Ground and 1st Floor Levels

No.	Load Case	Axial Load (kips)	Bending Moment (ft-kips)
	Dead (D)	188	-10.7
	Live (L)	59	-10.4
	Earthquake (E_h)	38	205.0
	Wind (W)	4	22.0
No.	**Load Combination**		
1	$1.4D + 1.7L$	364	-32.7
2	$0.75(1.4D + 1.7L + 1.7W)$	268	3.6
		278	-52.5
3	$0.9D + 1.3W$	164	19.0
		174	-38.2
4	$1.52D + 1.1L + 1.1E_h$	309	197.8
		393	-253.2
5	$0.79D + 1.1E_h$	107	217.0
		190	-234.0

Based on the load combinations in Table 4-14, a 22 × 22 in. column with 8-No. 9 bars (ρ_g = 1.65%) is adequate for column B2 below the first floor level. Figure 4-11, obtained from *PCACOL* (Ref. 4.3), shows the interaction diagram for this column. Also, 8-No. 8 bars (ρ_g = 1.31%) is adequate for column A2 below the first floor level (see Fig. 4-12). Note that the provided reinforcement ratios for both columns are within the allowable range of 1% and 6% (1921.4.3.1).

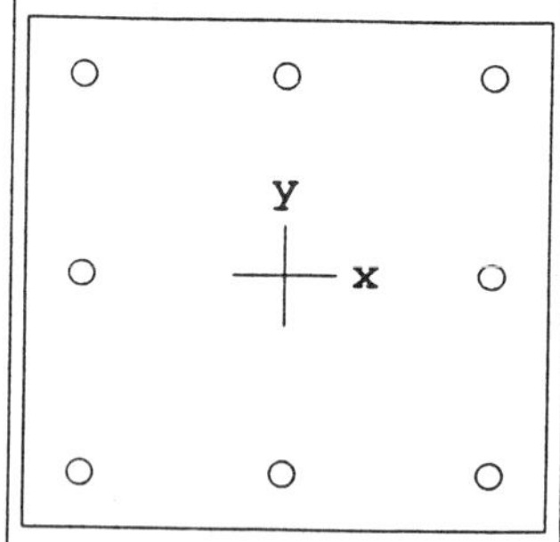

22.00 x 22.00 in

8-#9 at 1.65%

As = 8.00 in^2

Cover = 1.88 in

Spacing = 7.43 in

Ag = 484 in^2

Ix = 19521 in^4

Iy = 19521 in^4

Xo = 0.00 in

Yo = 0.00 in

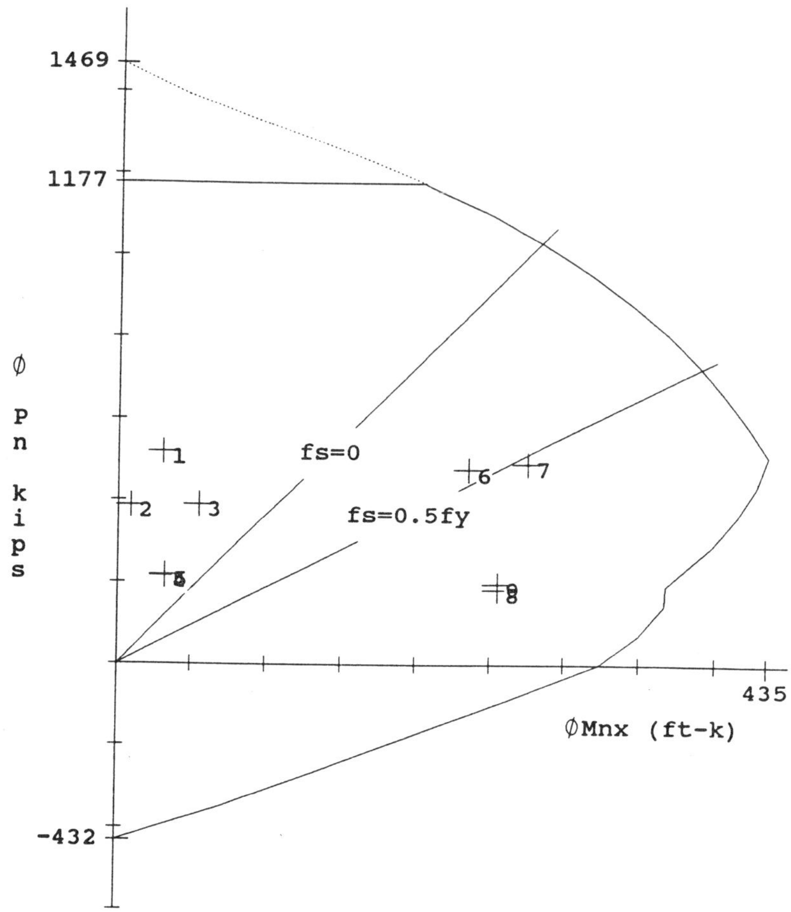

Licensed To: Portland Cement Association, Skokie, IL

File name: C:\DATA\COL\EB117IB.COL

Project: EB117

Column Id: Column B2

Engineer:

Date: Time:

Code:

Units: English

f'c = 4.00 ksi Ec = 3834 ksi

fc = 3.40 ksi eu = 0.003 in/in

Block stress profile, Beta1 = 0.85

fy = 60 ksi Es = 29000 ksi

Confinement: Tied

phi(c) = 0.70, phi(b) = 0.90, a = 0.80

X-axis slenderness is not considered.

Figure 4-11 Design Strength Interaction Diagram for Column B2 below the First Floor Level

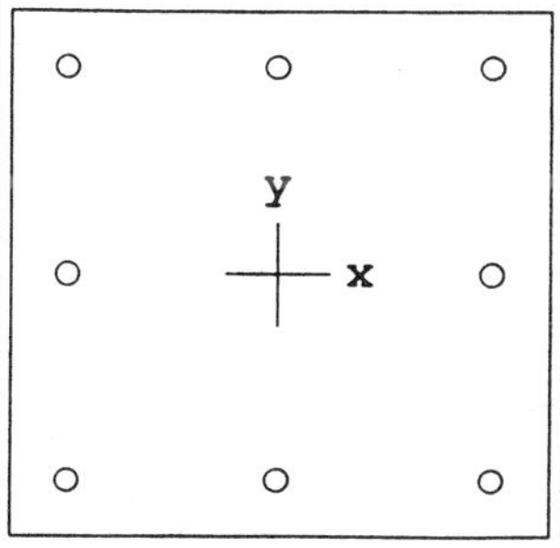

22.00 x 22.00 in

8-#8 at 1.31%

As = 6.32 in^2

Cover = 1.88 in

Spacing = 7.63 in

Ag = 484 in^2

Ix = 19521 in^4

Iy = 19521 in^4

Xo = 0.00 in

Yo = 0.00 in

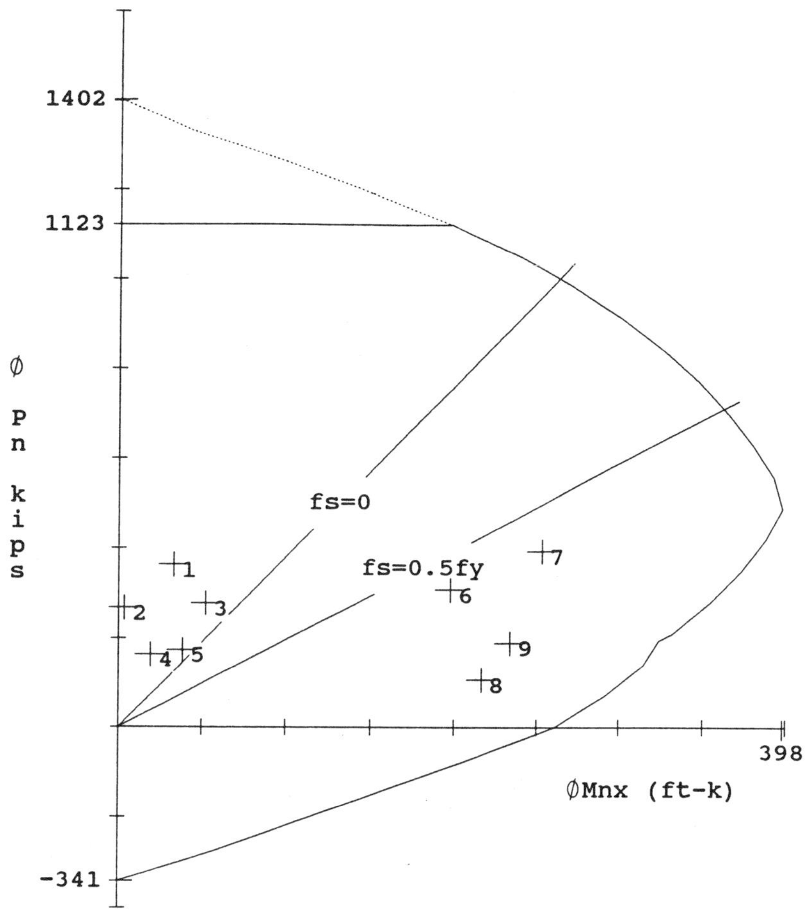

Licensed To: Portland Cement Association, Skokie, IL

File name: C:\DATA\COL\EB117EB.COL

Project: EB117		f'c = 4.00 ksi	Ec = 3834 ksi
Column Id: Column A2		fc = 3.40 ksi	eu = 0.003 in/in
Engineer:		Block stress profile, Beta1 = 0.85	
Date:	Time:	fy = 60 ksi	Es = 29000 ksi
Code:		Confinement: Tied	
Units: English		phi(c) = 0.70, phi(b) = 0.90, a = 0.80	

X-axis slenderness is not considered.

Figure 4-12 Design Strength Interaction Diagram for Column A2 below the First Floor Level

Relative Flexural Strengths of Columns and Girders.

Section 1921.4.2 requires that the sum of the flexural strengths of the columns at a joint must be greater than or equal to 6/5 the sum of the flexural strengths of the girders framing into that joint. Note that the column flexural strength is to be determined for the factored axial force resulting in the lowest flexural strength, consistent with the direction of the lateral forces considered. Only the seismic load combinations need to be considered when checking the relative strengths of the columns and girders.

At interior column B2, the negative design flexural strength of the beam is 328.0 ft-kips (see Table 4-12). The positive design flexural strength of the beam on the opposite face of the column is 265.0 ft-kips. For the upper column framing into the joint, the minimum flexural strength (ϕM_n) is approximately 362.5 ft-kips, which corresponds to a factored axial load of 123 kips (see Table 4-13 and pt. 8 in Fig. 4-13). Similarly, the minimum flexural strength is about 368.9 ft-kips for the lower column framing into the joint; this corresponds to a factored axial load of 184 kips (see pt. 8 in Fig. 4-11). Therefore,

$$\Sigma M_e = 362.5 + 368.9 = 731.4 \text{ ft-kips}$$

$$\Sigma M_g = 328.0 + 265.0 = 593.0 \text{ ft-kips}$$

$$731.4 \text{ ft-kips} > \frac{6}{5} \times 593.0 = 711.6 \text{ ft-kips} \quad \text{O.K.}$$

For exterior column A2, the negative design flexural strength of the beam is 265.0 ft-kips (see Table 4-12). For the upper column framing into the joint, the design flexural strength corresponding to a factored axial load of 77 kips is 294.4 ft-kips (see pt. 8 in Fig. 4-14). For the lower column, the design flexural strength corresponding to a factored axial load of 107 kips is 305.3 ft-kips (see pt. 8 in Fig. 4-12). Therefore,

$$294.4 + 305.3 = 599.7 \text{ ft-kips} > \frac{6}{5} \times 265.0 = 318.0 \text{ ft-kips} \quad \text{O.K.}$$

Confinement Reinforcement.

Special transverse reinforcement for confinement is required over a distance ℓ_o at the column ends where ℓ_o equals the maximum of (1921.4.4.4):

Depth of member = 22 in.

Clear height/6 = [(14 × 12) – 28]/6 = 23.3 in. (governs)

18 in.

According to 1921.4.4.2, the maximum allowable spacing of rectangular hoops within the 23.3 in. length shall not exceed one-fourth of the minimum member dimension which is equal to 0.25 × 22 = 5.5 in. or 4 in. (governs).

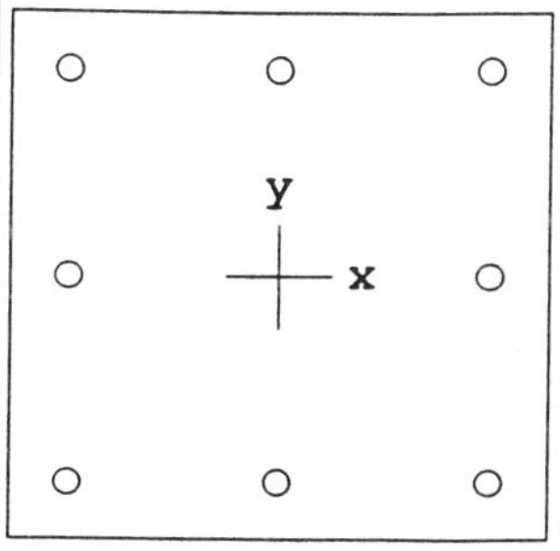

22.00 x 22.00 in

8-#9 at 1.65%
As = 8.00 in^2
Cover = 1.88 in
Spacing = 7.43 in

Ag = 484 in^2
Ix = 19521 in^4
Iy = 19521 in^4

Xo = 0.00 in
Yo = 0.00 in

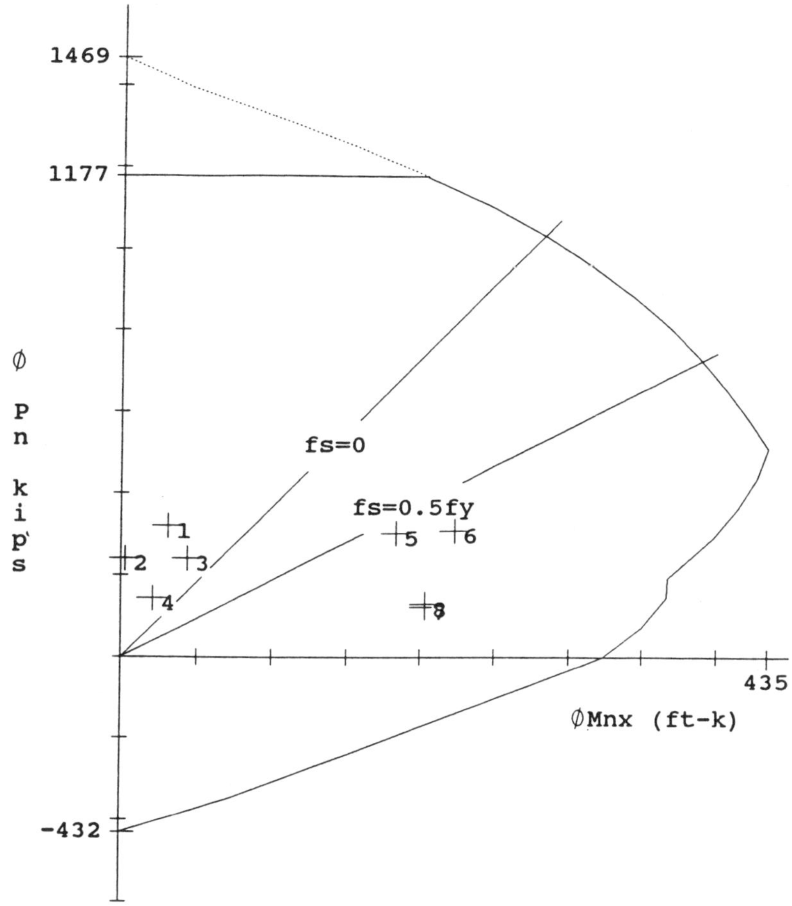

Licensed To: Portland Cement Association, Skokie, IL

File name: C:\DATA\COL\EB117IA.COL

Project: EB117	f'c = 4.00 ksi Ec = 3834 ksi
Column Id: Column B2	fc = 3.40 ksi eu = 0.003 in/in
Engineer:	Block stress profile, Beta1 = 0.85
Date: Time:	fy = 60 ksi Es = 29000 ksi
Code:	Confinement: Tied
Units: English	phi(c) = 0.70, phi(b) = 0.90, a = 0.80

X-axis slenderness is not considered.

Figure 4-13 Design Strength Interaction Diagram for Column B2 between the First and Second Floor Levels

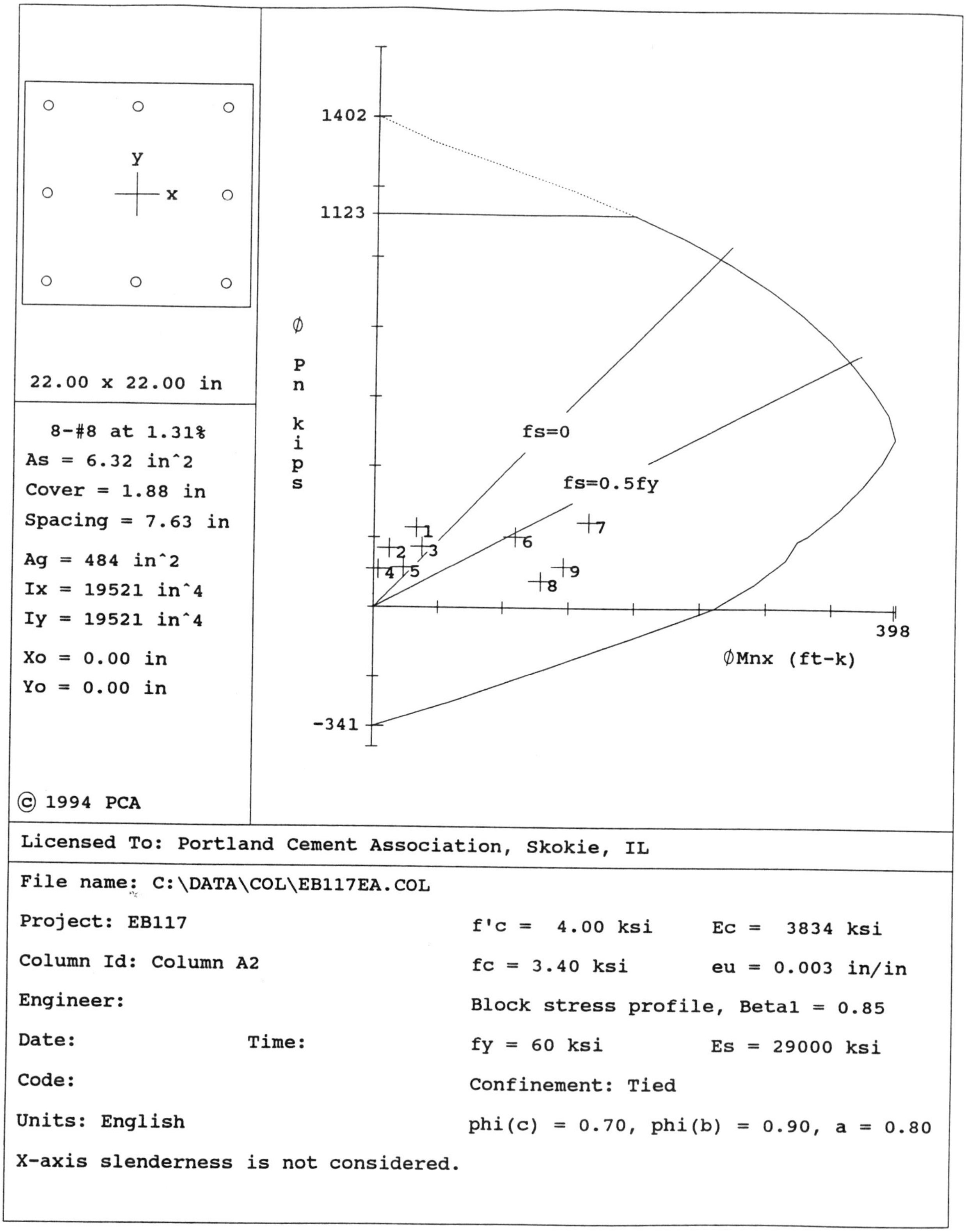

Figure 4-14 Design Strength Interaction Diagram for Column A2 between the First and Second Floor Levels

The minimum required cross-sectional area of hoop reinforcement (A_{sh}) is the larger value obtained from the following two equations (1921.4.4.1):

$$A_{sh} = \frac{0.3 s h_c f_c'}{f_{yh}} \left[\left(\frac{A_g}{A_{ch}} \right) - 1 \right]$$

$$= \frac{0.09 s h_c f_c'}{f_{yh}}$$

where s = spacing of transverse reinforcement

h_c = cross-sectional dimension of column core, measured center-to center of confining reinforcement

A_g = gross area of section

A_{ch} = cross-sectional area of member measured out-to-out of transverse reinforcement

f_{yh} = specified yield strength of transverse reinforcement

Using a hoop spacing of 4 in., f_{yh} = 60,000 psi, a clear cover of 1.5 in., and tentatively assuming No. 4 hoops, the required cross-sectional area of hoop reinforcement is:

$$A_{sh} = \frac{0.3 \times 4 \times \left[22 - (2 \times 1.5) - 0.5\right] \times 4}{60} \times \left[\frac{22^2}{(22-3)^2} - 1 \right] = 0.50 \text{ in.}^2 \quad \text{(governs)}$$

$$= \frac{0.09 \times 4 \times \left[22 - (2 \times 1.5) - 0.5\right] \times 4}{60} = 0.44 \text{ in.}^2$$

Note that one crosstie is required since the center-to-center distance between the outer hoop legs is 22 - (2 × 1.5) - 0.5 = 18.5 in. which is greater than 14 in. (1921.4.4.3). Using No. 4 hoops with one crosstie provides $A_{sh} = 3 \times 0.2 = 0.60$ in.^{2} which is greater than 0.50 in.^{2} Therefore, use a 4 in. spacing for the No. 4 hoops at the column ends.

Transverse Reinforcement for Shear.

Similar to beams, the design shear force for columns shall be determined considering the maximum forces that can be generated at the faces of the joints at each end (1921.4.5.1). These joint forces shall be determined using the probable moment strengths (M_{pr}) associated with the range of factored axial loads on the member; M_{pr} is calculated with the strength reduction factors equal to 1.0 and assuming that the stress in the tensile reinforcement is equal to $1.25 f_y$. The member shear forces need not exceed those determined from joint strengths based on the probable moment strength of the transverse members framing into the joint. It is important to note that the calculated shear force must not be taken less than the factored shear force obtained from analysis.

Figure 4-15 is the interaction diagram for column B2 with the strength reduction factors equal to 1.0 and f_y equal to 1.25 × 60 = 75 ksi. Conservatively assume that the maximum moment (M_{pr}) is equal to the moment at the balanced point which is approximately 654 ft-kips. Thus, the design shear force based on the probable moment strength at the column ends is equal to (2 × 654)/[14 - (28/12)] = 112.1 kips.

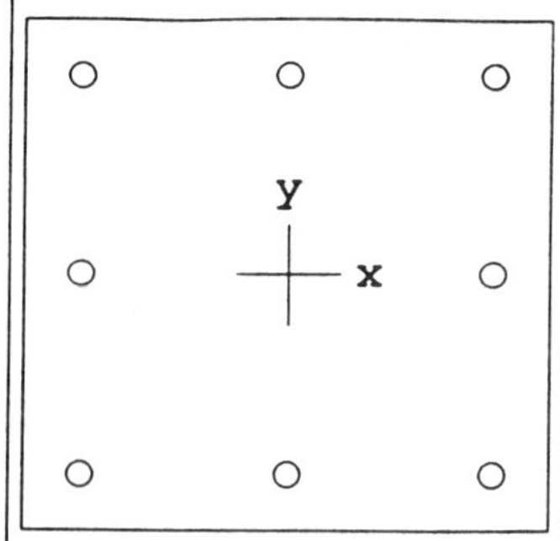

22.00 x 22.00 in

8-#9 at 1.65%
As = 8.00 in^2
Cover = 1.88 in
Spacing = 7.43 in

Ag = 484 in^2
Ix = 19521 in^4
Iy = 19521 in^4

Xo = 0.00 in
Yo = 0.00 in

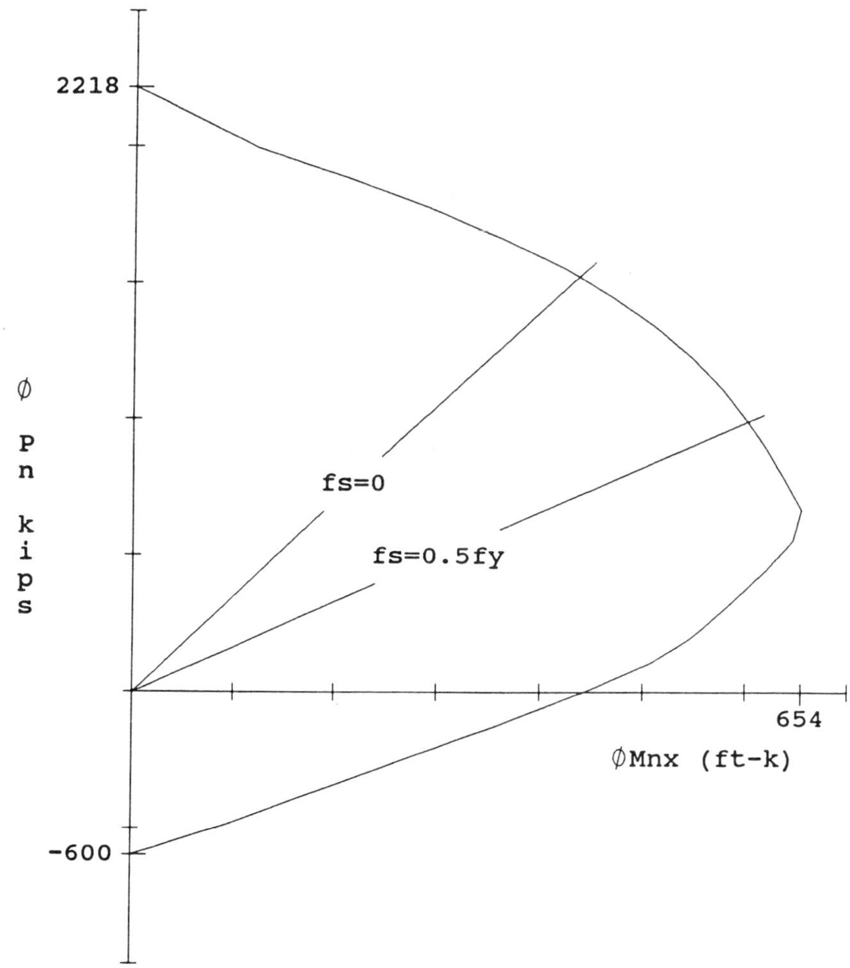

File name: C:\DATA\COL\EB117IB.COL

Project: EB117	f'c = 4.00 ksi Ec = 3834 ksi
Column Id: Column B2	fc = 3.40 ksi eu = 0.003 in/in
Engineer:	Block stress profile, Beta1 = 0.85
Date: Time:	fy = 75 ksi Es = 29000 ksi
Code:	Confinement: Other
Units: English	phi(c) = 1.00, phi(b) = 1.00, a = 1.00

X-axis slenderness is not considered.

Figure 4-15 Design Strength Interaction Diagram for Column B2 with f_y = 75 ksi and ϕ = 1.0

The positive probable moment strength of the beam framing into the joint at the face of column B2 at the first floor level is 364.5 ft-kips (see Fig. 4-8). The negative probable moment strength of the beam on the other side of the column is 450.0 ft-kips. Assuming that the bending moments at the top and bottom of the column supporting the first floor level are equal, the design shear force based on the probable moment strengths of the beams is:

$$V_u = \frac{(M_{pr}^+ + M_{pr}^-)}{\ell_1} = \frac{(364.5 + 450.0)}{14 - \frac{28}{12}} = 69.8 \text{ kips}$$

where ℓ_1 is the clear length of the column in the first story. This factored shear force is greater than the factored shear force determined from analysis. Since the factored axial forces for column B2 are greater than $A_g f'_c /20 = (22^2 \times 4)/20 = 96.8$ kips, the shear strength of the concrete may be used (1921.4.5.2 and 1911.3.1.2):

$$V_c = 2\left(1 + \frac{N_u}{2{,}000A_g}\right)\sqrt{f'_c}\, b_w d = 2 \times \left(1 + \frac{184{,}000}{2{,}000 \times 22^2}\right) \times \sqrt{4{,}000} \times 22 \times 19.4/1{,}000 = 64.3 \text{ kips}$$

$$V_s = \frac{A_v f_y d}{s} = \frac{(3 \times 0.2) \times 60 \times 19.4}{4} = 174.6 \text{ kips}$$

$$\phi(V_c + V_s) = 0.85(64.3 + 174.6) = 203.0 \text{ kips} > 69.8 \text{ kips} \quad \text{O.K.}$$

Thus, the No. 4 hoops at 4 in. on center required for confinement over the distance ℓ_o at the column ends is also adequate for the transverse reinforcement required for shear.

The remainder of the column length must contain hoop reinforcement with center-to-center spacing not to exceed either six times the diameter of the longitudinal column bars which is equal to 6 × 1.128 = 6.8 in. or 6 in. (1921.4.4.6). In this case, the 6 in. spacing controls.

The shear strength requirements for column A2 can be checked in a similar fashion.

Splice Length for Column Vertical Bars.

The lap splice length of the No. 9 bars in column B2 can be determined according to 1912.2.3:

$$\frac{\ell_d}{d_b} = \frac{3}{40}\frac{f_y}{\sqrt{f'_c}}\frac{\alpha\beta\gamma\lambda}{\left(\frac{c + K_{tr}}{d_b}\right)}$$

where α = reinforcement location factor = 1.0 (no top bar effect)

β = coating factor = 1.0 (uncoated reinforcement)

γ = reinforcement size factor = 1.0 (No. 7 or larger bar size)

λ = lightweight aggregate concrete factor = 1.0 (normal weight concrete)

$$c = \text{spacing or cover dimension} = \begin{cases} 1.5 + 0.5 + \dfrac{1.128}{2} = 2.6 \text{ in. (governs)} \\ \dfrac{22 - 2(1.5 + 0.5) - 1.128}{2 \times 2} = 4.2 \text{ in.} \end{cases}$$

$$K_{tr} = \text{transverse reinforcement index} = \frac{A_{tr} f_{yt}}{1{,}500 sn} = \frac{3 \times 0.2 \times 60{,}000}{1{,}500 \times 4 \times 3} = 2.0$$

$$\frac{c + K_{tr}}{d_b} = \frac{2.6 + 2.0}{1.128} = 4.1 > 2.5, \text{ use } 2.5$$

$$\frac{\ell_d}{d_b} = \frac{3}{40} \times \frac{60{,}000}{\sqrt{4{,}000}} \times \frac{1.0 \times 1.0 \times 1.0 \times 1.0}{2.5} = 28.5$$

$$\ell_d = 28.5 \times 1.128 = 32.1 \text{ in.} = 2.7 \text{ ft}$$

According to 1921.4.3.2, the lap splices must be located within the center half of the column length and must be proportioned as tension splices. From Fig. 4-11 it can be seen that the stress in the longitudinal bars is greater than $0.5f_y$ for the load combinations corresponding to the effects of the seismic forces. Thus, a Class B splice is required (1912.17.2.3). The Class B splice length is equal to $1.3 \times 2.7 = 3.5$ ft. Reinforcement details for column B2 are shown in Fig. 4-16.

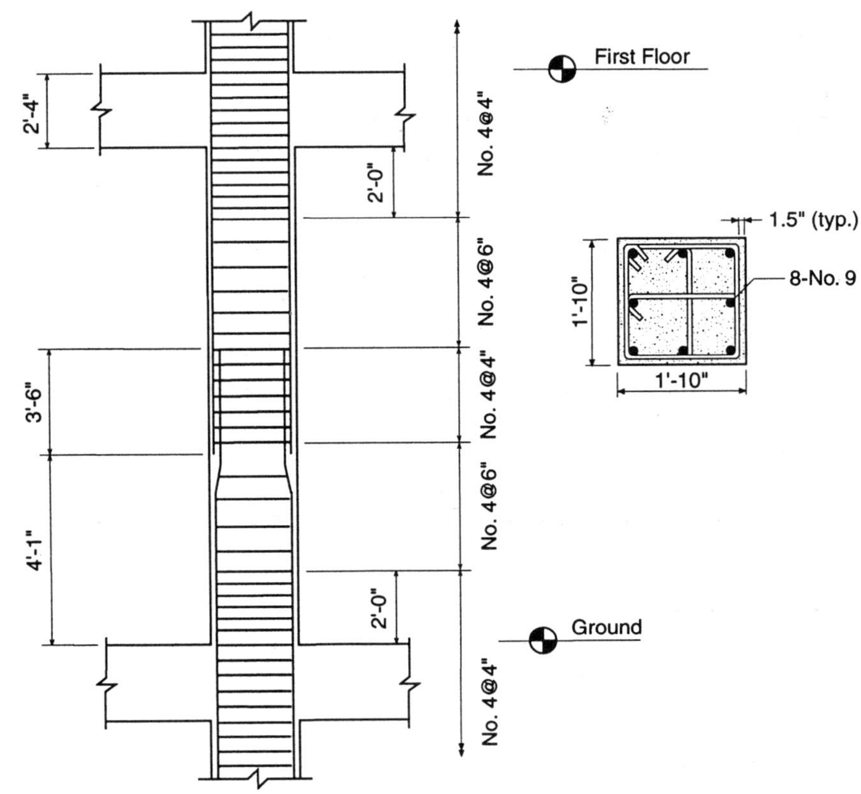

Figure 4-16 Reinforcement Details for Column B2

4.6.4 Design of Beam-Column Joint

Interior Joint.

Section 1921.5 requires that transverse hoop reinforcement as specified in 1921.4.4 shall be provided within the joint, unless the joint is confined by structural members as specified in 1921.5.2.2. For an interior joint, beams that have the same width as the column frame into all four sides of the joint. Thus, within the joint depth, transverse reinforcement equal to at least one-half the amount required by 1921.4.4.1 shall be provided, and the spacing is permitted to be $0.25 \times 22 = 5.5$ in. (governs) or 6 in. For detailing simplicity, use the transverse reinforcement of the column (No. 4 @ 4 in.) through the joint.

Figure 4-17 shows the interior beam-column joint at the first floor level. The shear strength is checked in the N-S direction along column line 2. The shear force at section *x-x* is obtained by subtracting the column horizontal shear force from the sum of the tensile force in the top beam reinforcement and the compressive force at the top of the beam on the opposite face of the column.

The column horizontal shear force (V_h) can be obtained by assuming that the adjoining floors are deformed so that plastic hinges form at their junctions with the columns, with M^-_{pr} (beam) = 450.0 ft-kips and M^+_{pr} (beam)= 364.5 ft-kips (1921.5.1.1).

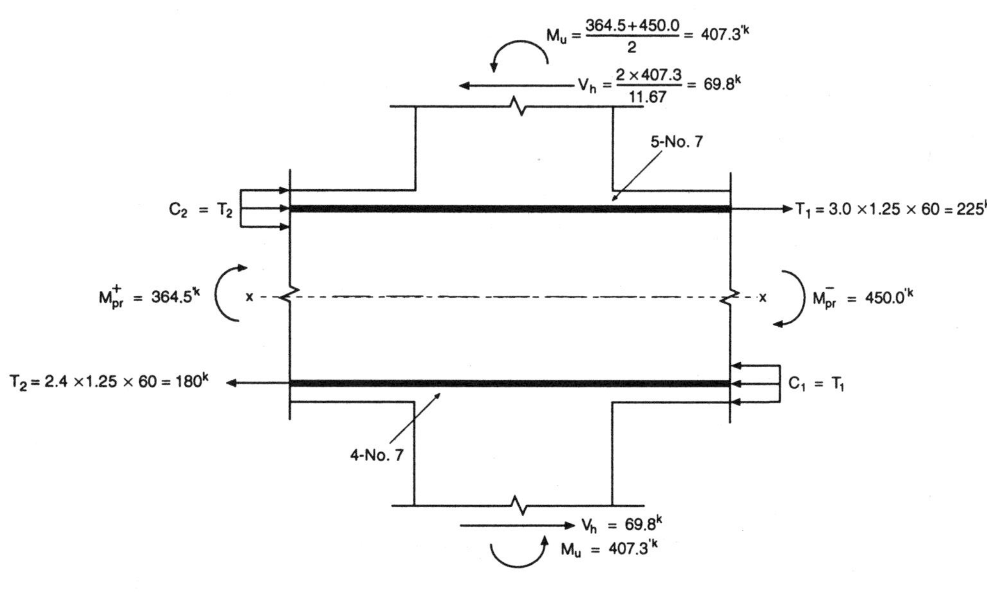

Figure 4-17 Shear Analysis of Interior Beam-Column Joint Along Column Line 2

It is assumed that the beam moments are resisted by the columns above and below the joint inversely proportional to their lengths. Since both columns have the same length, the column moment is equal to (450.0 + 364.5)/2 = 407.3 ft-kips. The horizontal shear force (V_h) at the column ends is:

$$V_h = \frac{2 \times 407.3}{14 - \frac{28}{12}} = 69.8 \text{ kips}$$

The tensile force in the top reinforcement (5-No. 7 bars) is equal to (5 × 0.60) × 1.25 × 60 = 225.0 kips. The compressive force on the opposite side of the column is equal to the tensile force of 4-No. 7 bars which is (4 × 0.60) × 1.25 × 60 = 180.0 kips.

The net shear at section x-x is $T_1 + C_2 - V_h$ = 225.0 + 180.0 - 69.8 = 335.2 kips.

For a joint confined on all four faces, the nominal shear strength (ϕV_c) is (1921.5.3.1):

$$\phi V_c = \phi 20 \sqrt{f'_c} A_j$$

where A_j is equal to the effective cross-sectional area within a joint in a plane parallel to the plane of the reinforcement generating shear in the joint. The joint depth is the overall depth of the column. Since the beams are as wide as the column, the effective width of the joint is equal to the column width. Thus, A_j = 22 × 22 = 484 in.^{2} and:

$$\phi V_c = 0.85 \times 20 \times \sqrt{4,000} \times 484/1,000 = 520.4 \text{ kips} > 335.2 \text{ kips} \quad \text{O.K.}$$

Exterior Joint.

The shear strength will be checked in the N-S direction for the exterior beam connection along column line 2. The edge column has beams framing on three sides; thus, the joint is not confined by framing members according to 1921.5.2.2 so that the transverse reinforcement required for the column must be continued through the joint.

Figure 4-18 shows a section through an exterior beam-column joint at the first floor level. The shear force at section x-x is obtained by subtracting the column horizontal shear force from the tensile force in the top beam reinforcement.

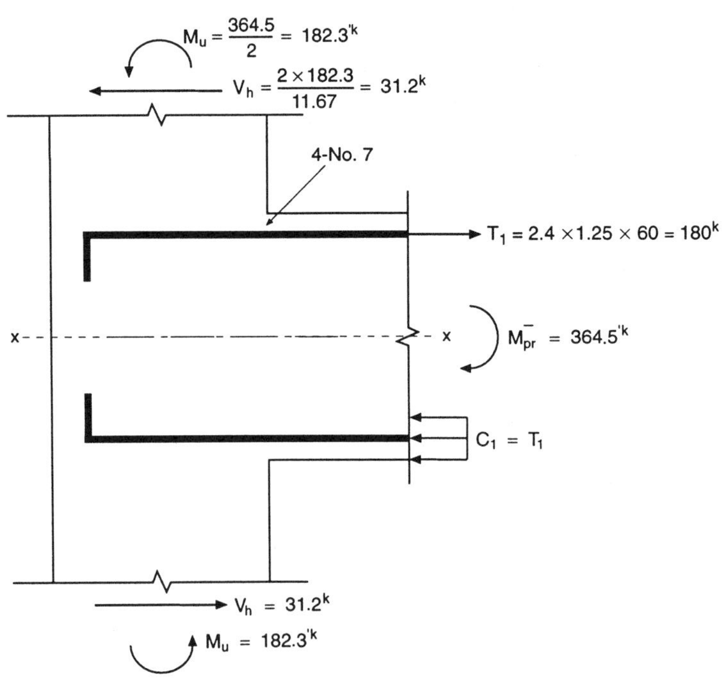

Figure 4-18 Shear Analysis of Exterior Beam-Column Joint Along Column Line 2

The column horizontal shear force (V_h) can be obtained by assuming that the beam in the adjoining floor is deformed so that a plastic hinge form at the junction with the column, with M_{pr}^{-}(beam) = 364.5 ft-kips.

Since the columns above and below the joint have the same length, the column moment is equal to 364.5/2 = 182.3 ft-kips. The horizontal shear force (V_h) is:

$$V_h = \frac{2 \times 182.3}{14 - \frac{28}{12}} = 31.2 \text{ kips}$$

The tensile force in the top reinforcement (4-No. 7 bars) is equal to (4 × 0.60) × 1.25 × 60 = 180.0 kips.

The net shear at section *x-x* is T_1 - V_h = 180.0 - 31.2 = 148.8 kips.

For a joint confined on all three faces, the nominal shear strength (ϕV_c) is (1921.5.3.1):

$$\phi V_c = \phi 15 \sqrt{f_c'} A_j = 0.85 \times 15 \times \sqrt{4,000} \times 22^2 / 1,000 = 390.3 \text{ kips} > 148.8 \text{ kips} \qquad \text{O.K.}$$

4.7 References

4.1 *PCA-Frame – Three-Dimensional Static Analysis of Structures*, Portland Cement Association, Skokie, IL, 1992.

4.2 *Notes on ACI 318-95, Building Code Requirements for Structural Concrete with Design Applications*, Portland Cement Association, Skokie, IL, 1996.

4.3 *PCACOL – Strength Design of Reinforced Concrete Column Sections*, Portland Cement Association, Skokie, IL, 1994.

5 BUILDING FRAME SYSTEM

5.1 INTRODUCTION

A typical plan and elevation of a 10-story reinforced concrete shearwall-frame building is shown in Fig. 5-1. The computation of the seismic design forces according to the 1997 edition of the Uniform Building Code (UBC) is illustrated below. Typical beams, columns, and shearwalls are designed and detailed for the combined effects of gravity, seismic, and wind forces.

In a building frame system, the shearwalls are designed to carry the total lateral loads, and the frames provide support for the gravity loads. However, since the shearwalls and beam-column frames form one structural system and will laterally deflect by virtually the same amount, it is important to ensure that compatibility requirements be addressed in the design of the components so that the beam-column frames do not collapse under large lateral seismic deflections.

It is assumed mainly for simplicity that the columns and shearwalls have constant cross-sections throughout the height of the building, and that the bases of the lowest story segments are fixed. The beams and joists are also assumed to have the same dimensions at all floor levels. Although the member dimensions in this example are within the practical range, the structure itself is a hypothetical one, and has been chosen mainly for illustrative purposes.

5.2 DESIGN DATA

- Material Properties

  Concrete: f'_c = 4,000 psi, w_c = 145 pcf (beams, joists, and columns)

  f'_c = 6,000 psi, w_c = 145 pcf (shearwalls, including boundary elements)

  Reinforcement: f_y = 60,000 psi

- Service Loads

  Live loads: roof = 20 psf
  floor = 50 psf (Table 16-A)

  Weight of joists: 109 psf

  Superimposed dead load: 25 psf

- Seismic Design Data

  Seismic Zone 4 (Z = 0.40; see Table 16-I)
  Seismic Source Type B (Table 16-U), with the site located 5 km from a known seismic source
  Soil Profile Type S_B (Table 16-J)

  For standard occupancy, Seismic Importance Factor (I) = 1.0 (Table 16-K)
  For the building frame system R = 5.5 (Table 16-N)

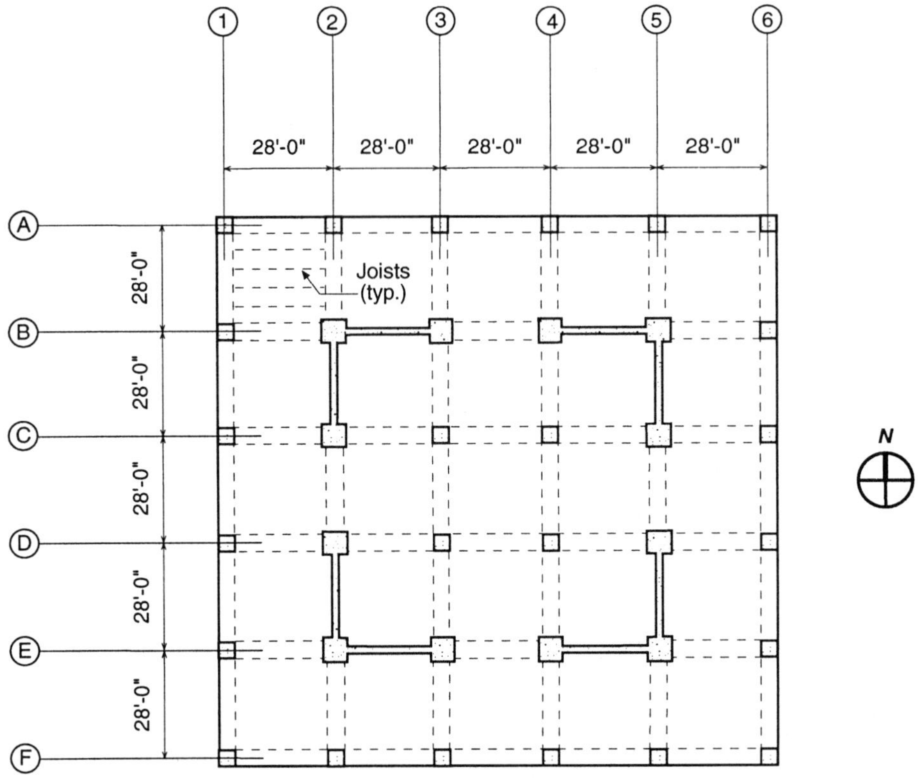

Beams: 24 x 24.5 in.

Columns: 24 x 24 in.

Shearwalls: boundary elements: 36 x 36 in.

wall thickness: 12 in.

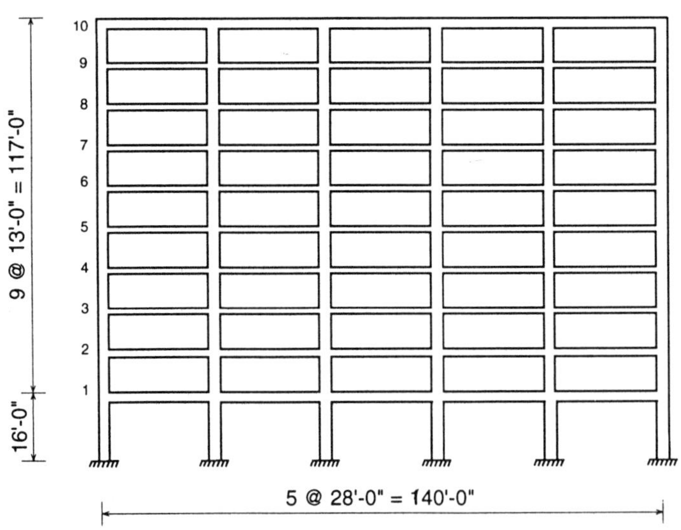

Figure 5-1 Typical Floor Plan and Elevation of Example Building

- Wind Design Data

  Basic wind speed = 70 mph
  Exposure C
  For standard occupancy, Wind Importance Factor (I_w) = 1.0 (Table 16-K)

5.3 SEISMIC LOAD ANALYSIS

5.3.1 Seismic Forces

The seismic base shear (V) is determined below according to the static force procedure (1630.2). Note that this procedure is applicable for this regular building which is less than 240 ft in height (1629.8.3), and that the magnitude of V is the same in both the N-S and E-W directions. The base shear is distributed vertically to each floor level according to 1630.5.

Seismic Dead Load (*W*).

The dead loads for the roof level, a typical floor, and the first floor level are contained in Table 5-1. The total seismic dead load of the building (W) is equal to 3,516 + (8 × 3,896) + 3,984 = 38,668 kips.

Table 5-1 Dead Loads at Various Floor Levels (kips)

	Horizontal Elements*	Vertical Elements	Total
Roof	3,135	381	3,516
Typical Floor	3,135	761	3,896
Floor Level 1	3,135	849	3,984

**includes the weight of the beams, joists, and the superimposed dead load*

Approximate Period (*T*).

Method A in 1630.2.2 is used to compute the period (T). For the building frame system:

Building height (h_n) = 133 ft

Numerical coefficient $C_t = 0.020$

Period $T = C_t\,(h_n)^{3/4} = 0.020 \times (133)^{3/4} = 0.783$ sec

Base Shear (*V*).

The total design base shear (V) is determined from Eq. (30-4) using $T = 0.783$ sec:

$$V = \frac{C_v I W}{R T}$$

where $C_v = 0.40 N_v = 0.40 \times 1.2 = 0.48$ for Seismic Zone 4, soil profile type S_B, and seismic source type B with the site located 5 km from the seismic source (see Tables 16-R and 16-T).

Therefore,

$$V = \frac{0.48 \times 1.0 \times 38{,}668}{5.5 \times 0.783} = 4{,}308 \text{ kips}$$

The total design base shear need not exceed the value from Eq. (30-5):

$$V = \frac{2.5C_a IW}{R} = \frac{2.5 \times 0.40 \times 1.0 \times 38,668}{5.5} = 7,031 \text{ kips}$$

where $C_a = 0.40N_a = 0.40 \times 1.0 = 0.40$ for Seismic Zone 4, soil profile type S_B, and seismic source type B with the site located 5 km from the seismic source (see Tables 16-Q and 16-S).

Also, V shall not be less than the following values from Eqs. (30-6) and (30-7), the later equation being applicable for Seismic Zone 4:

$$V = \begin{cases} 0.11C_a IW = 0.11 \times 0.40 \times 1.0 \times 38,668 = 1,701 \text{ kips} \\ \dfrac{0.8ZN_v IW}{R} = \dfrac{0.8 \times 0.40 \times 1.2 \times 1.0 \times 38,668}{5.5} = 2,700 \text{ kips} \end{cases}$$

Thus, use V = 4,308 kips in both the N-S and E-W directions.

Lateral Forces (F_x).

The total base shear (V) is distributed over the height of the building in conformance with Eqs. (30-13), (30-14), and (30-15). The lateral force (F_x) at each floor level and the story shears (V_x) are contained in Table 5-2.

Table 5-2 Seismic Forces and Story Shears

Level	Story weight, w_x (kips)	Height, h_x (ft)	$w_x h_x$	Lateral Force, F_x* (kips)	Story Shear, V_x (kips)
10	3,516	133	467,628	903	903
9	3,896	120	467,520	667	1,570
8	3,896	107	416,872	595	2,165
7	3,896	94	366,224	523	2,688
6	3,896	81	315,576	450	3,138
5	3,896	68	264,928	378	3,516
4	3,896	55	214,280	306	3,822
3	3,896	42	163,632	234	4,056
2	3,896	29	112,984	161	4,217
1	3,984	16	63,744	91	4,308
Σ	38,668		2,853,388		

$$^*F_x = \frac{(V - F_t) w_x h_x}{\sum_{i=1}^{n} w_i h_i}$$ *(see 1630.5).*

Note that at the top of the building, the required concentrated force F_t is equal to $0.07TV = 0.07 \times 0.783 \times 4{,}308 = 236$ kips which is less than $0.25V = 1{,}077$ kips (1630.5). This force must be added to the force F_n to obtain the total force at the roof level:

$$\text{Total force at roof level} = F_n + F_t = \frac{(V - F_t)w_n h_n}{\sum_{i=1}^{n} w_i h_i} + F_t = \frac{(4{,}308 - 236) \times 467{,}628}{2{,}853{,}388} + 236 = 903 \text{ kips}$$

5.3.2 Analysis Results

The member reactions resulting from the seismic forces contained in Table 5-2 were determined in the following manner. A lateral load analysis was performed on the portion of the structure that is the lateral force resisting (LFR) system. For this building, the LFR system consists of the shearwalls and the coupling beams framing into the walls. It was assumed that the wall segments and coupling beams parallel to the direction of analysis resisted all of the effects of the lateral loads in that direction. Also, 50 percent of the gross section properties of the walls and beams were used for stiffness considerations in accordance with 1633.2.4. To account for accidental torsion, the mass at each level was assumed to be displaced from the calculated center of mass a distance equal to 5 percent of the building dimension at that level perpendicular to the direction of the force under consideration (1630.6). Table 5-3 contains the results from the structural analysis for seismic forces acting on the LFR system.

Table 5-3 Results of Analysis of Lateral Force Resisting System Due to Seismic Forces

Shearwall D2-E2				
Story	Axial Load (kips)	Bending Moment (ft-kips) Top	Bending Moment (ft-kips) Bottom	Shear Force (kips)
10	34	956	2,469	264
9	71	-1,441	7,394	458
8	106	-6,415	14,629	632
7	141	-13,642	23,843	785
6	174	-22,914	34,826	916
5	205	-33,945	47,291	1,027
4	233	-46,515	61,025	1,116
3	257	-60,357	75,751	1,184
2	274	-75,259	91,278	1,232
1	286	-90,951	111,052	1,256

Coupling Beam C2-D2		
Level	Bending Moment (ft-kips)	Shear Force (kips)
10	-478.5	34.2
9	-513.7	36.7
8	-489.1	34.9
7	-493.1	35.2
6	-463.4	33.1
5	-439.3	31.4
4	-387.0	27.6
3	-333.3	23.8
2	-246.6	17.6
1	-163.4	11.7

Table 5-4 contains the displacements (Δ_S) obtained from the static, elastic analysis of the LFR system using the design seismic forces, including accidental torsional effects. The table also contains the maximum inelastic response displacements (Δ_M) which are computed from Eq. (30-17):

$$\Delta_M = 0.7R\Delta_S$$

where R is equal to 5.5. The interstory drifts from the corresponding Δ_S and Δ_M are also contained in the table.

Table 5-4 Lateral Displacements and Interstory Drifts Due to Seismic Forces (in.)

Story	Δ_S	Drift	Δ_M	Drift
10	3.73	0.48	14.36	1.85
9	3.25	0.46	12.51	1.77
8	2.79	0.47	10.74	1.81
7	2.32	0.46	8.93	1.77
6	1.86	0.44	7.16	1.69
5	1.42	0.40	5.47	1.54
4	1.02	0.36	3.93	1.39
3	0.66	0.30	2.54	1.15
2	0.36	0.22	1.39	0.85
1	0.14	0.14	0.54	0.54

All framing elements not required by design to be part of the LFR system must be designed and shown to be adequate for vertical load carrying capacity when subjected to the expected deformations caused by the seismic forces (1633.2.4). The expected deformations are the greater of the maximum inelastic response displacements (Δ_M) or the deformations induced by a story drift of 0.0025 times the story height. Note that P-Δ effects on these elements must also be taken into account.

The analysis of the non-lateral-force-resisting system was performed using the maximum inelastic response displacements (Δ_M) at all floor levels, since these deformations are greater than those induced by a story drift of 0.0025 times the story height (1633.2.4). P-Δ effects were also included in the analysis and were found to be negligible. Table 5-5 contains a summary of the results of the analysis for a column and beam accounting for deformation compatibility and P-Δ effects.

5.3.3 Story Drift Limitation and P-Δ Effects

The elements of the LFR system must meet the story drift limitations of 1630.10, and must be designed for P-Δ effects if applicable (1630.1.3). According to 1630.10.2, the calculated story drifts using Δ_M shall not exceed 0.020 times the story height for structures having a fundamental period of 0.7 seconds or greater. For the 13 ft story height, the maximum drift is equal to $0.020 \times 13 \times 12 = 3.12$ in., and for the 16 ft story height it is equal to 3.84 in. It is clear from Table 5-4 that for all stories, the lateral drifts obtained from the prescribed lateral forces are less than the limiting values.

According to 1630.1.3, P-Δ effects need not be considered for Seismic Zones 3 and 4 when the calculated story drift using Δ_S is less than or equal to $0.02h_s/R$, which is $0.02 \times 16 \times 12/5.5 = 0.70$ in. for the first floor and $0.02 \times 13 \times 12/5.5 = 0.57$ in. for a typical floor. It can be seen from Table 5-4 that P-Δ effects need not be considered for the elements of the LFR system at any story level.

Table 5-5 Results of Analysis of Non-Lateral-Force-Resisting System Accounting for Deformation Compatibility and P-Δ Effects

Column A3				
Story	Axial Load (kips)	Bending Moment (ft-kips)		Shear Force (kips)
		Top	Bottom	
10	42.8	626.2	-491.7	86.0
9	95.5	260.1	-306.7	43.6
8	145.6	409.6	-392.9	61.7
7	196.8	338.0	-378.6	55.1
6	246.1	327.3	-399.5	55.9
5	292.3	260.0	-351.2	47.0
4	334.3	249.4	-363.7	47.2
3	371.1	162.2	-333.5	38.1
2	400.0	79.5	-262.7	26.3
1	420.1	25.9	-373.1	24.9

Beam A3-B3			
Level	Bending Moment (ft-kips)		Shear Force (kips)
	Exterior	Interior	
10	-626.2	572.1	42.8
9	-751.8	725.0	52.7
8	-716.3	686.1	50.1
7	-730.9	700.5	51.1
6	-705.9	676.7	49.4
5	-659.5	632.4	46.1
4	-600.6	576.0	42.0
3	-525.9	504.1	36.8
2	-413.0	396.7	28.9
1	-288.6	274.6	20.1

5.4 WIND LOAD ANALYSIS

Table 5-6 contains the wind loads computed according to 1609 with the basic wind speed equal to 70 mph, exposure C, and a wind importance factor (I_w) equal to 1.0.

Table 5-6 Design Wind Forces (for entire structure)

Level	Height h_x (ft)	Tributary Height (ft)	Windward			Leeward			Total Force (kips)
			C_e	Pressure* (psf)	Force** (kips)	C_e	Pressure* (psf)	Force** (kips)	
10	133	6.5	1.71	17.2	15.9	1.71	10.8	10.0	25.9
9	120	13.0	1.67	16.9	31.1	1.71	10.8	19.9	51.0
8	107	13.0	1.63	16.4	30.3	1.71	10.8	19.9	50.2
7	94	13.0	1.58	16.0	29.4	1.71	10.8	19.9	49.3
6	81	13.0	1.53	15.4	28.5	1.71	10.8	19.9	48.4
5	68	13.0	1.47	14.9	27.4	1.71	10.8	19.9	47.3
4	55	13.0	1.41	14.2	26.2	1.71	10.8	19.9	46.1
3	42	13.0	1.33	13.4	24.8	1.71	10.8	19.9	44.7
2	29	13.0	1.22	12.3	22.7	1.71	10.8	19.9	42.6
1	16	14.5	1.08	10.8	22.4	1.71	10.8	22.2	44.6

**Windward pressure = $C_e C_q q_s I_w = C_e \times 0.8 \times 12.6 \times 1.0 = 10.08C_e$ psf*

Leeward pressure = $C_e C_q q_s I_w = 1.71 \times 0.5 \times 12.6 \times 1.0 = 10.8$ psf

***Force = pressure × tributary height × 142.0/1,000*

It is clear from Table 5-6 that the wind forces which act on the building are considerably smaller than the seismic forces (see Table 5-2). Thus, the effects of the wind loads on the members will not be considered in the remainder of this example.

5.5 Gravity Load Analysis

5.5.1 Beams

The gravity load bending moments are determined for a beam that is not part of the LFR system (beam A3-B3) and one that is part of the LFR system (beam C2-D2). The gravity loads acting on beam A3-B3 are as follows:

$$w_{DL} = [(109+25)\times 28] + \left(\frac{24\times 24.5}{144}\times 145\right) = 4{,}344 \text{ plf}$$

$$w_{LL} = 50\times 28 = 1{,}400 \text{ plf}$$

The coefficients in 1908.3.3 can be used to compute the bending moments at the faces of the supports since all of the conditions given in 1908.3.3 are satisfied in this case. Table 5-7 contains the dead load and live load moments for beam A3-B3.

Table 5-7 Service Dead and Live Load Bending Moments (ft-kips) for Beam A3-B3 on a Typical Floor

	Moment*	M_{DL}	M_{LL}
Ext. Negative	$w\ell_n^2/16$	-176.6	-56.9
Positive	$w\ell_n^2/14$	201.8	65.0
Int. Negative	$w\ell_n^2/10$	-282.5	-91.0

**see 1908.3.3*

Beam C2-D2, which is part of the LFR system, frames into shearwalls at both ends. Both ends of the beam can be considered to be fixed, with a bending moment equal to $w\ell_n^2/12$ at the face of the support. Note that the gravity loads on this beam are the same as those on beam A3-B3. The dead and live load bending moments for beam C2-D2 are given in Table 5-8.

Table 5-8 Service Dead and Live Load Moments (ft-kips) for Beam C2-D2 on a Typical Floor

	Moment	M_{DL}	M_{LL}
Negative	$w\ell_n^2/12$	-226.3	-72.9
Positive	$w\ell_n^2/24$	113.1	36.5

5.5.2 Columns and Shearwalls

The service axial loads for column A3 and shearwall D2-E2 are contained in Tables 5-9 and 5-10, respectively. Note that column A3 is not part of the LFR system. In both cases, the live load reduction factor was computed in accordance with 1607.5.

Table 5-9 Service Dead and Live Load Axial Forces for Column A3

Story	Dead Load* (psf)	Live Load (psf)	Supported Area (ft^2)	Reduced Live Load** (psf)	Live Load (kips)	Cumulative Dead Load† (kips)	Cumulative Live Load (kips)
10	134	20	420	15.68	7	82	7
9	134	50	840	22.40	9	164	16
8	134	50	1,260	20.00	8	246	24
7	134	50	1,680	20.00	8	328	32
6	134	50	2,100	20.00	8	410	40
5	134	50	2,520	20.00	8	492	48
4	134	50	2,940	20.00	8	574	56
3	134	50	3,360	20.00	8	656	64
2	134	50	3,780	20.00	8	738	72
1	134	50	4,200	20.00	8	822	80

**includes the weight of the joists and the superimposed dead load*
***see 1607.5*
†includes beam stem weight of 0.48 kips/ft and column weight of 0.58 kips/ft

Table 5-10 Service Dead and Live Load Axial Forces for Shearwall D2-E2

Story	Dead Load* (psf)	Live Load (psf)	Supported Area (ft^2)	Reduced Live Load** (psf)	Live Load (kips)	Cumulative Dead Load† (kips)	Cumulative Live Load (kips)
10	134	20	1,568	12.00	19	322	19
9	134	50	3,136	20.00	31	644	50
8	134	50	4,704	20.00	31	966	81
7	134	50	6,272	20.00	31	1,288	112
6	134	50	7,840	20.00	31	1,610	143
5	134	50	9,408	20.00	31	1,932	174
4	134	50	10,976	20.00	31	2,254	205
3	134	50	12,544	20.00	31	2,576	236
2	134	50	14,112	20.00	31	2,898	267
1	134	50	15,680	20.00	31	3,238	298

**includes the weight of the joists and the superimposed dead load*
***see 1607.5*
†includes beam stem weight of 0.48 kips/ft and wall weight of 6.2 kips/ft

5.6 Design for Combined Load Effects

5.6.1 Load Combinations

The basic load combinations for the strength design method are given in 1612.2.1. According to the exceptions at the end of that section, the load combinations of 1909.2 are applicable for concrete where the load combinations do not include seismic forces. Also, the load combinations of 1612.2.1 which include seismic forces must be multiplied by 1.1 for concrete (see Sect. 2.2 for discussion on the 1.1 factor). Thus, the following load combinations are applicable to concrete members which are part of the LFR system subjected to the combined effects of dead (D), live (L), wind (W), and earthquake (E) forces:

(1) $1.4D + 1.7L$

(2) $0.75(1.4D + 1.7L + 1.7W)$

(3) $0.9D + 1.3W$

(4) $1.1(1.2D + f_1L + 1.0E) = 1.32D + 1.1f_1L + 1.1E$

(5) $1.1(0.9D + 1.0E) = 0.99D + 1.1E$

where f_1 = 1.0 for floors in places of public assembly, for live loads in excess of 100 psf, and for garage live loads

= 0.5 for other live loads

The earthquake force (E) is determined from Eq. (30-1):

$$E = \rho E_h + E_v$$

where E_h = earthquake force due to the base shear (V)

E_v = the load effect resulting from the vertical component of the earthquake motion

$= 0.5C_a ID = 0.5 \times 0.40 \times 1.0 \times D = 0.2D$ in this case

ρ = reliability/redundancy factor (see Eq. 30-3):

$$1.0 \leq \rho = 2 - \frac{20}{r_{max}\sqrt{A_B}} \leq 1.5$$

r_{max} = maximum element-story shear ratio

A_B = ground floor area of the structure, ft^2

According to 1630.1.1, the maximum element-story shear ratio (r_{max}) is defined as the largest of the element-story shear ratios (r_i) which occurs in any of the story levels at or below the two-thirds height level of the building.

For shearwalls, r_i is the maximum value of the wall shear multiplied by $10/\ell_w$ and divided by the total story shear, where ℓ_w is the length of the wall in feet. Table 5-11 contains the maximum shear forces in the shearwall due to seismic forces, which include the effects from accidental torsional (see Table 5-3), the story shear (see Table 5-2), and the values of r_i for the story levels below the two-thirds height level of the building. As expected, the element-story shear ratio (r_i) is a constant over the height of the building, so that r_{max} is equal to 0.094. Therefore,

$$\rho = 2 - \frac{20}{0.094\sqrt{142.0^2}} = 0.5 < 1.0$$

According to 1630.1.1, ρ shall not be taken less than one. Thus, with ρ set equal to 1, the seismic force (E) is:

$$E = E_h + 0.2D$$

Substituting E and $f_1 = 0.5$ (live load < 100 psf) into load combinations 4 and 5 above results in the following (see Sect. 2.4.1 regarding the use of $0.2D$ in Eq. (5)):

(4) $1.32D + (1.1 \times 0.5)L + 1.1E = 1.32D + 0.55L + 1.1(E_h + 0.2D) = 1.54D + 0.55L + 1.1E_h$

(5) $0.99D + 1.1E = 0.99D + 1.1(E_h - 0.2D) = 0.77D + 1.1E_h$

The values of E_h can be found in Table 5-3 for the elements of the LFR system.

Table 5-11 Element-Story Shear Ratios (r_i) Due to Seismic Forces

Story	Wall Shear (kips)	Story Shear (kips)	r_i
6	916	3,138	0.094
5	1,207	3,516	0.094
4	1,116	3,822	0.094
3	1,184	4,056	0.094
2	1,232	4,217	0.094
1	1,256	4,308	0.094

For members that are not part of the LFR system, the forces on the members obtained from the application of the expected deformations may be considered as factored forces (1633.2.4). The following load combinations are applicable to the design of members that are not part of the LFR system (see 1921.7):

(1) $1.4D + 1.4L + E$

(2) $0.9D + E$

where the values of E are given in Table 5-5 for column A3 and beam A3-B3.

Note that the special seismic load combinations of 1612.4 are not applicable in this example.

5.6.2 Design of Beam A3-B3

Design for Flexure.

A summary of the bending moments at the first floor and third floor levels, based on the applicable load combinations, is given in Table 5-12 for beam A3-B3. The gravity load moments are obtained from Table 5-7 and the compatibility moments are taken from Table 5-5.

Table 5-12 Summary of Design Bending Moments (ft-kips) for Beam A3-B3

No.	Load Case		First Floor	Third Floor
	Dead (*D*)	Ext. Neg.	-176.6	-176.6
		Positive	201.8	201.8
		Int. Neg.	-282.5	-282.5
	Live (*L*)	Ext. Neg.	-56.9	-56.9
		Positive	65.0	65.0
		Int. Neg.	-91.0	-91.0
	Earthquake (*E*)	Ext. Neg.	±288.6	±525.9
		Positive	—	—
		Int. Neg.	±274.6	±504.1
No.	Load Combination			
1	$1.4D + 1.4L + E$	Ext. Neg.	-38.3	199.0
			-615.5	-852.8
		Positive	373.5	373.5
		Int. Neg.	-797.5	-1,027.0
			-248.3	-18.8
2	$0.9D + E$	Ext. Neg.	129.7	367.0
			-447.5	-684.8
		Positive	181.6	181.6
		Int. Neg.	-528.9	-758.4
			20.4	249.9

At the first floor level, the maximum bending moment is 797.5 ft-kips at the face of the first interior support, and 9.59 in.^{2} of reinforcing steel is required at this location. This amount of reinforcement is less than the maximum amount permitted ($0.025bd$) which is 13.20 in.^{2} (1921.3.2.1). At the same location at the third floor level, the bending moment is 1,027 ft-kips, and 13.35 in.^{2} of reinforcing steel is required, which is greater than the maximum allowable value. Since the magnitude of the seismic bending moment at this location increases at the floor levels above level 3 (see Table 5-5), the required amount of reinforcement also increases, resulting in areas of steel which are also greater than the maximum allowable value. One way to circumvent this problem is to provide flexural reinforcement for the factored bending moments due to gravity loads only and to satisfy the provisions of 1921.7.3, since, based on this smaller amount of reinforcement, the design moment strength of the member (ϕM_n) would be less than the total combined moment due to lateral displacement and factored gravity loads (M_u). Table 5-13 contains the required reinforcement for the factored gravity load moments and the design moment strengths (ϕM_n) for beam A3-B3.

As was noted above, the provisions of 1921.7.3 must be satisfied for beam A3-B3. In particular, the materials used in this example satisfy the provisions of 1921.2.4 and 1921.2.5 (1921.7.3.1). Also, since the factored gravity axial force on this member is negligible, the flexural reinforcement requirements of 1921.3.2.1 must be satisfied (1921.7.3.2). The prescribed minimum and maximum amounts of reinforcement are given at the bottom of Table 5-13. It is clear that the provided areas of steel fall well within the allowable range. In order to satisfy the minimum reinforcement requirements of 1921.3.2.1, 3-No. 8 bars (A_s = 2.37 in.^{2} > 1.76 in.^{2}) must be continuous at both the top and bottom of the member; this automatically satisfies the minimum continuous bar requirements specified in 1921.3.2.1 as well.

Table 5-13 Summary of Required Reinforcement for Beam A3-B3 due to Factored Gravity Loads

Load Combination	Location	M_u (ft-kips)	A_s* (in.^{2})	Reinforcement	ϕM_n (ft-kips)
$1.4D + 1.7L$	Ext. Neg.	-344.0	3.70	5-No. 8	365.2
	Positive	393.0	4.28	6-No. 8	432.1
	Int. Neg.	-550.2	6.20	8-No. 8	559.6

**Max. $A_s = 0.025 \times 24 \times 22 = 13.20$ in.^{2} (1921.3.2.1)*

Min. $A_s = 3\sqrt{4,000} \times 24 \times 22/60,000 = 1.67$ in.^{2}

$= 200 \times 24 \times 22/60,000 = 1.76$ in.^{2} (governs)

Design for Shear.

The shear provisions of 1921.3.4 must be satisfied for beam A3-B3 (1921.7.3.2). Shear forces are determined assuming that moments of opposite sign corresponding to the probable moment strength (M_{pr}) act at the joint faces and that the member is loaded with the tributary gravity load along its span. It is important to note that sidesway to the right and to the left must both be considered when calculating the maximum design shear forces.

The probable moment strength (M_{pr}) is determined at a section using the stress in the tensile reinforcement equal to $1.25f_y$ and a strength reduction factor (ϕ) equal to 1.0. The following equation can be used to compute M_{pr} :

$$M_{pr} = A_s(1.25f_y)\left(d - \frac{a}{2}\right)$$

where $$a = \frac{A_s(1.25f_y)}{0.85f'_c\,b}$$

For example, for sidesway to the right, the interior joint must be subjected to the negative moment M_{pr} which is determined as follows:

For 8-No. 8 top bars, $A_s = 8 \times 0.79 = 6.32$ in.^{2}

$$a = \frac{A_s(1.25f_y)}{0.85f'_c\,b} = \frac{6.32 \times 1.25 \times 60}{0.85 \times 4 \times 24} = 5.81 \text{ in.}$$

$$M_{pr} = A_s(1.25f_y)\left(d - \frac{a}{2}\right) = 6.32 \times 1.25 \times 60 \times \left(22 - \frac{5.81}{2}\right) = 9{,}051 \text{ in.-kips} = 754.3 \text{ ft-kips}$$

Similarly, for the exterior joint, the positive moment M_{pr} based on the 6-No. 8 bottom bars is equal to 587.2 ft-kips. The probable moment strengths for sidesway to the left can be obtained in a similar fashion.

The gravity load on the beam is:

$$w = [(109 + 25) \times 28] + \left(\frac{24 \times 24.5}{144} \times 145\right) + (50 \times 28) = 5{,}744 \text{ plf}$$

Figure 5-2 shows the exterior beam A3-B3 and the probable moment strengths (M_{pr}) at the joint faces for sidesway to the right and to the left. This figure also shows the tributary gravity loads on the beam, as well as the combined shear forces at the faces of the columns.

In general, the shear strength is assumed to be provided by both concrete (V_c) and reinforcing steel in the form of hoops or stirrups (V_s). According to 1921.3.4.2, when proportioning the transverse reinforcement over the length $2h = 2 \times 24.5 = 49$ in. = 4.1 ft from the face of the supports (1921.3.3.1), the shear strength of the concrete (V_c) is to be set equal to zero if the earthquake-induced shear force is 50% or more of the total shear and the factored axial compressive force is less than $A_g f'_c/20$, where A_g is the gross area of the beam. As noted above, the axial force on this member is negligible, and the maximum earthquake-induced shear force at the face of the interior support is (754.3 + 587.2)/25.5 = 52.6 kips which is less than 0.5 × 125.8 = 62.9 kips. However, at sections located more than approximately 3.6 ft from the face of the interior support, the earthquake-induced shear force is greater than one-half of the total design shear. Thus, V_c is set equal to zero at every section located within 4.1 ft from the face of the interior support. The maximum shear force V_s is (1911.1):

$$\phi V_s = V_u - \phi V_c$$

$$V_s = \frac{V_u}{\phi} - 0 = \frac{125.8}{0.85} = 148.0 \text{ kips}$$

Shear strength contributed by shear reinforcement shall not exceed $(V_s)_{max}$ (1911.5.6.8):

$$(V_s)_{max} = 8\sqrt{f_c'}\, b_w d = 8\sqrt{4{,}000} \times 24 \times 22/1{,}000 = 267.2 \text{ kips} > 148.0 \text{ kips} \quad \text{O.K.}$$

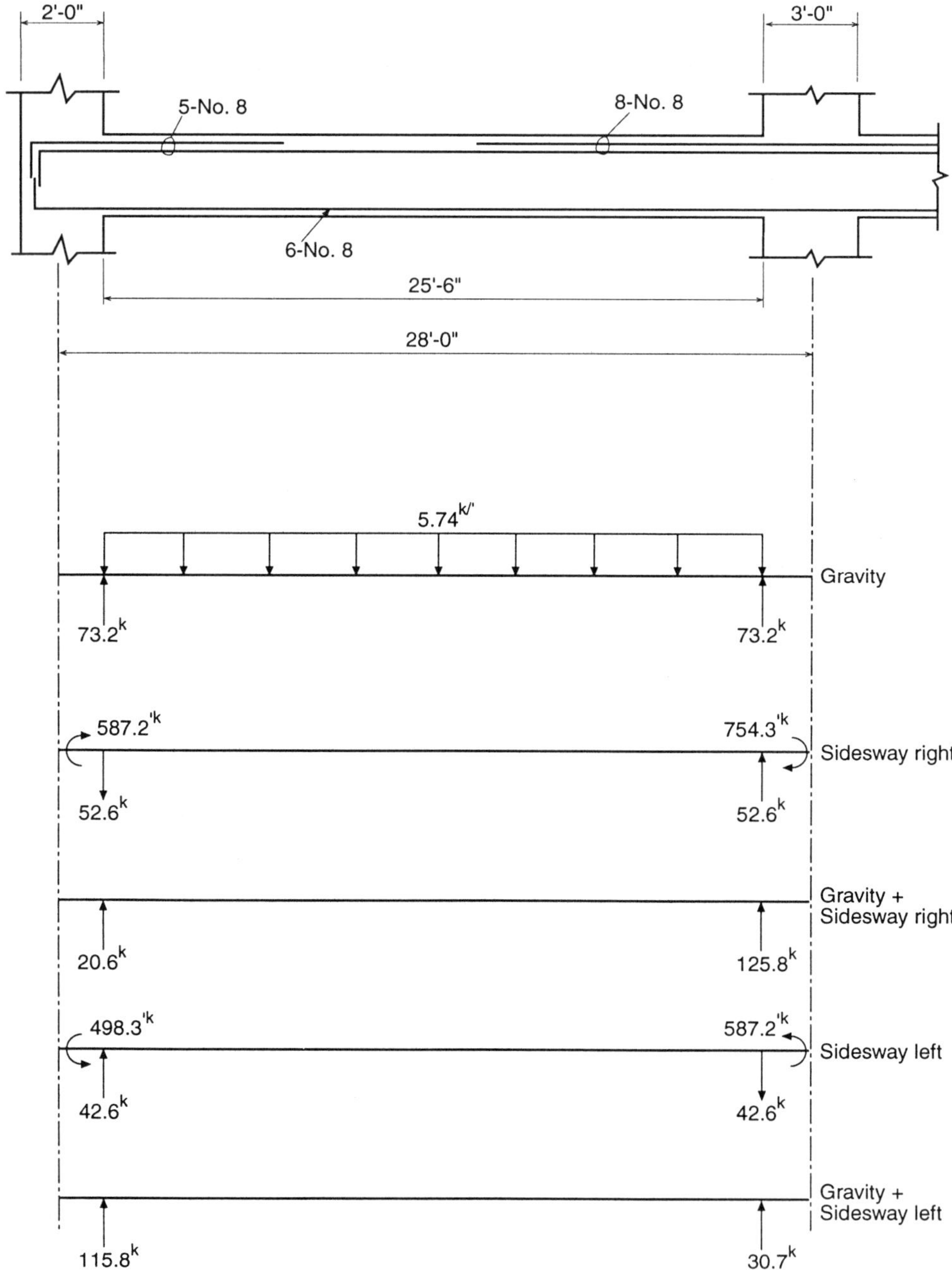

Figure 5-2 Design Shear Forces for Beam A3-B3

However, V_s is greater than $4\sqrt{f_c'}\, b_w d = 133.6$ kips so that the maximum spacings specified in 1911.5.4.1 must be reduced by one-half (1911.5.4.3).

Required spacing of No. 4 closed stirrups (hoops) with 4 legs for a factored shear force of 148.0 kips is (1911.5.6.2):

$$s = \frac{A_v f_y d}{V_s} = \frac{(4 \times 0.2) \times 60 \times 22}{148.0} = 7.1 \text{ in.}$$

Maximum allowable hoop spacing (s_{max}) from the face of the support is the smallest of the following (1911.5.4.1, 1911.5.4.3):

$$s_{max} = \frac{1}{2} \times \frac{d}{2} = \frac{22}{4} = 5.5 \text{ in. (governs)}$$

$$= \frac{1}{2} \times 24 = 12.0 \text{ in.}$$

Therefore, hoops must be spaced at 5 in. on center with the first one located at 2 in. from the face of the support. Eleven hoops are to be placed at this spacing. Where hoops are no longer required, stirrups may be used and may be proportioned including the nominal shear strength provided by the concrete (V_c).

At a distance of 52 in. from the face of the interior support, V_u = 101.0 kips. The maximum shear force V_s is:

$$V_s = \frac{V_u}{\phi} - V_c = \frac{101.0}{0.85} - \frac{2\sqrt{4,000} \times 24 \times 22}{1,000} = 52.0 \text{ kips} < 4\sqrt{f_c'}\, b_w d = 133.6 \text{ kips}$$

Required spacing of No. 4 stirrups with two legs for this factored shear force is:

$$s = \frac{A_v f_y d}{V_s} = \frac{(2 \times 0.2) \times 60 \times 22}{52.0} = 10.2 \text{ in.}$$

Use No. 4 stirrups spaced at 10 in. Note that this spacing is less than the maximum allowable spacing which is equal to $d/2$ = 11 in. (1921.7.3.2).

At the face of the exterior support, the maximum earthquake-induced shear force is (587.2 + 498.3)/25.5 = 42.6 kips which is less than 0.5 × 115.8 = 57.9 kips (see Fig. 5-2). The earthquake-induced shear force becomes greater than one-half of the total design shear at a distance approximately equal to 5.3 ft from the face of the exterior support. Note that this distance is greater than $2h$ = 4.1 ft. Thus, V_c need not be set equal to zero (1921.3.4.2). The maximum shear force V_s at the face of the exterior support is:

$$V_s = \frac{V_u}{\phi} - V_c = \frac{115.8}{0.85} - \frac{2\sqrt{4,000} \times 24 \times 22}{1,000} = 69.5 \text{ kips}$$

Required spacing of No. 4 hoops with two legs is:

$$s = \frac{A_v f_y d}{V_s} = \frac{(2 \times 0.2) \times 60 \times 22}{69.5} = 7.6 \text{ in.} < \frac{d}{2} = 11 \text{ in.}$$

Therefore, hoops must be spaced at 7 in. on center with the first one located at 2 in. from the face of the support. Eight hoops are to be placed at this spacing. For the remaining portion of the beam, use No. 4 stirrups (2 legs) spaced at 10 in.

Reinforcing Bar Cutoff Points.

The negative reinforcement at the interior support is 8-No. 8 bars. The location where five of the eight bars can be terminated will be determined.

The loading used to find the cutoff point of the 5-No. 8 bars at this location is 0.9 times the dead load in combination with the probable flexural strengths (M_{pr}) at the ends of the member; this combination will produce the longest bar lengths. The design flexural strength (ϕM_n) provided by 3-No. 8 bars is 225.3 ft-kips. Therefore, the 5 reinforcing bars can be terminated after the required moment strength (M_u) has been reduced to 225.3 ft-kips.

With $\phi = 1.0$ and $f_s = 75$ ksi, M_{pr} is equal to 754.3 ft-kips at the interior support and 587.2 ft-kips at the exterior support. The load is equal to $0.9 \times \{[(109 + 25) \times 28] + [(24 \times 24.5/144) \times 145]\}/1{,}000 = 3.91$ kips/ft. The distance from the face of the interior support to where the moment equals 225.3 ft-kips is obtained by summing moments about section *a-a* in Fig. 5-3:

$$3.91x\left(\frac{x}{2}\right) - 102.5x + 754.3 = 225.3$$

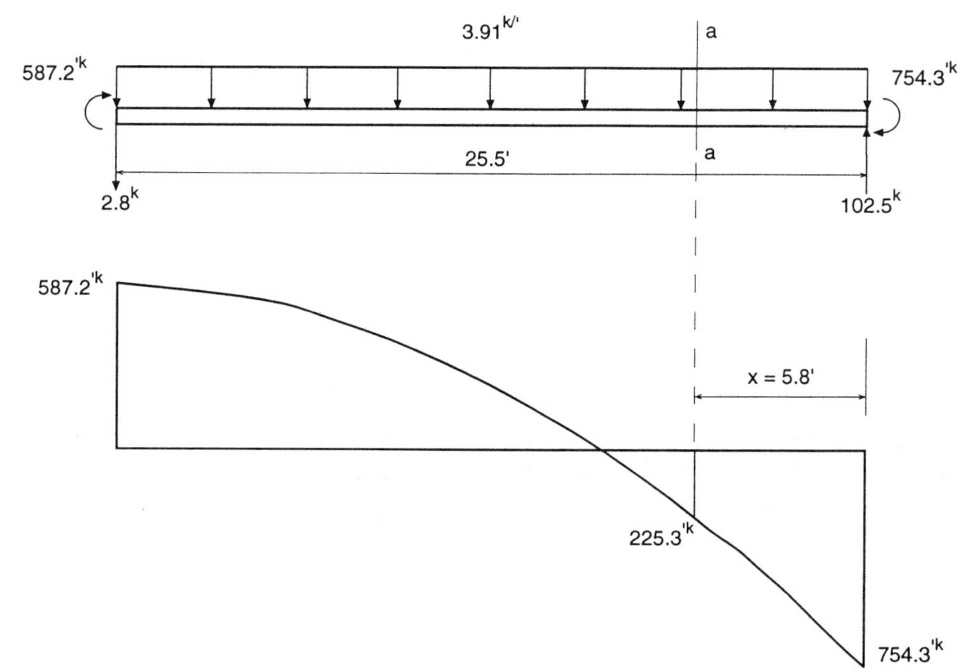

Figure 5-3 Moment Diagram for Cutoff Location of Negative Bars at Interior Support of Beam A3-B3

Solving for x gives a distance of 5.8 ft. The 5-No. 8 bars must extend a distance $d = 22$ in. or $12d_b = 12 \times 1.0 = 12.0$ in. beyond the distance x (1912.10.3). Thus, from the face of the support, the total bar length must be at least equal to $5.8 + (22/12) = 7.6$ ft. Also, the bars must extend a full development length ℓ_d beyond the face of the support (1912.10.4). The development length can be determined from the provisions of 1912.2.3:

$$\frac{\ell_d}{d_b} = \frac{3}{40} \frac{f_y}{\sqrt{f_c'}} \frac{\alpha\beta\gamma\lambda}{\left(\dfrac{c + K_{tr}}{d_b}\right)}$$

where α = reinforcement location factor = 1.3 (top bar)

β = coating factor = 1.0 (uncoated reinforcement)

γ = reinforcement size factor = 1.0 (No. 8 bar)

λ = lightweight aggregate concrete factor = 1.0 (normal weight concrete)

$$c = \text{spacing or cover dimension} = \begin{cases} 1.5 + 0.5 + \dfrac{1.0}{2} = 2.5 \text{ in.} \\ \dfrac{24 - 2(1.5 + 0.5) - 1.0}{2 \times 7} = 1.36 \text{ in. (governs)} \end{cases}$$

K_{tr} = transverse reinforcement index = 0 (conservative)

$$\frac{c + K_{tr}}{d_b} = \frac{1.36 + 0}{1.0} = 1.36$$

$$\frac{\ell_d}{d_b} = \frac{3}{40} \times \frac{60{,}000}{\sqrt{4{,}000}} \times \frac{1.3 \times 1.0 \times 1.0 \times 1.0}{1.36} = 68.0$$

$$\ell_d = 68.0 \times 1.0 = 68.0 \text{ in.} = 5.7 \text{ ft} < 7.6 \text{ ft}$$

Thus, the total required length of the 5-No. 8 bars must be at least 7.6 ft beyond the face of the support. It is important to note that flexural reinforcement shall not be terminated in a tension zone unless one or more of the conditions of 1912.10.5 is satisfied. In this case, the point of inflection is approximately 8.9 ft from the face of the right support which is greater than 7.6 ft. Therefore, the 5-No. 8 bars can not be terminated here unless one of the conditions of 1912.10.5 is satisfied. In lieu of increasing the shear reinforcement, use 1912.10.5.1, and check if the factored shear force (V_u) at the cutoff point does not exceed two-thirds of that permitted ($2\phi V_n/3$). For No. 4 stirrups at 10 in. on center that are provided in this region of the beam:

$$\phi V_n = \phi(V_s + V_c) = 0.85 \times \left(\frac{0.4 \times 60 \times 22}{10} + 66.8\right) = 101.7 \text{ kips}$$

$$\frac{2}{3}\phi V_n = 67.8 \text{ kips} < V_u = 72.6 \text{ kips at 7.6 ft from face of support}$$

Since V_u is greater than $2\phi V_n/3$, the total required length of the 5-No. 8 bars must be at least 8.9 ft beyond the face of the interior support.

The cutoff point for two of the 5-No. 8 bars at the exterior support can be determined in a similar fashion.

The required splice length of the top reinforcing bars can be determined using the provisions of 1912.15.1. The development length (ℓ_d) is 37.0 in. (1912.2.3). Thus, the Class A splice length is equal to 1.0 × 37.0 = 37.0 in. Use a 38 in. long splice.

Figure 5-4 shows the reinforcement details for beam A3-B3.

5.6.3 Design of Beam C2-D2

According to 1921.2.1.4, in Seismic Zones 3 and 4, all reinforced concrete members that are part of the lateral load resisting system must satisfy the requirements of 1921.2 through 1921.7, in addition to the requirements of 1901 through 1917. In particular, for coupling beams, the provisions of 1921.6.10 must be satisfied. Since ℓ_n/d = 25 × 12/22 = 13.6 > 4, the design of this member must conform to the requirements of 1921.2 and 1921.3. A check of the conditions of 1921.3.1 is shown below:

- Factored axial compressive force on beams is negligible. O.K.
- Minimum clear span = 25 ft > $4d$ = 4 × (24.5 – 2.5)/12 = 7.3 ft O.K.
- $\frac{\text{Width}}{\text{Depth}} = \frac{24}{24.5} = 0.98 > 0.3$ O.K.
- Width = 24 in. > 10 in. O.K.
- Width = 24 in. < width of supporting column + (1.5 × depth of beam)

  < 36 + (1.5 × 24.5) = 72.75 in. O.K.

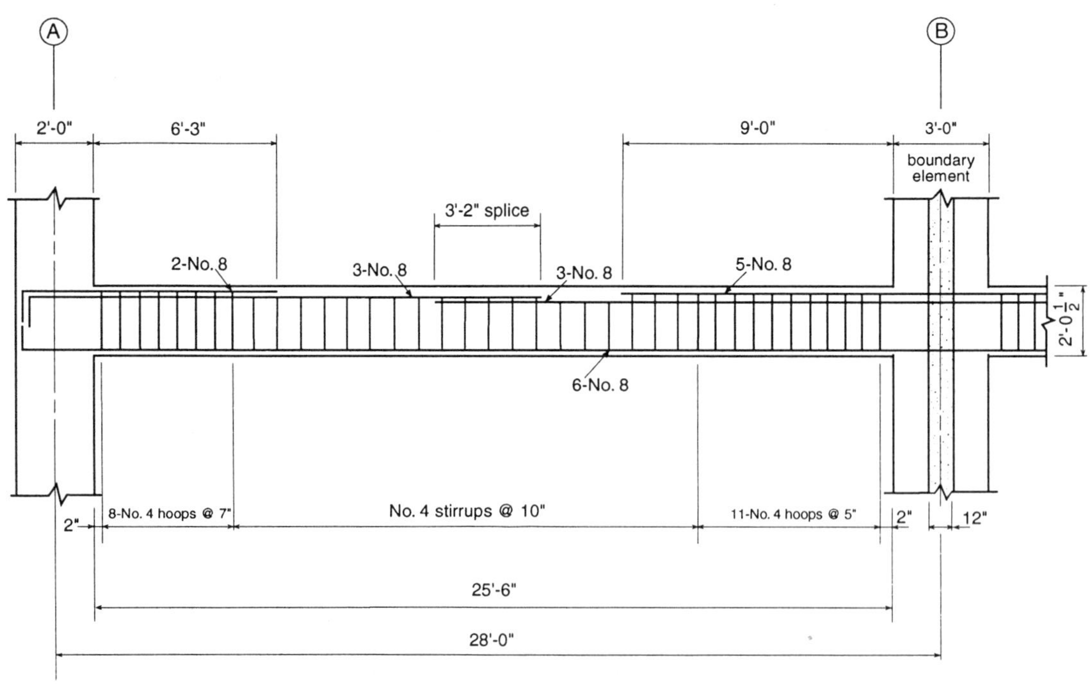

Figure 5-4 Reinforcement Details for Beam A3-B3

Design for Flexure.

A summary of the bending moments at the fourth floor level, based on the applicable load combinations, is given in Table 5-14 for beam C2-D2. The gravity load moments are obtained from Table 5-8 and the moments from the seismic load effects are taken from Table 5-3.

Table 5-14 Summary of Design Bending Moments (ft-kips) for Beam C2-D2 at Floor Level 4

	Load Case		Fourth Floor
	Dead (D)	Negative Positive	-226.3 113.1
	Live (L)	Negative Positive	-72.9 36.5
	Earthquake (E)	Negative Positive	±387.0 —
No.	Load Combination		
1	$1.4D + 1.7L$	Negative Positive	-440.8 220.4
4	$1.54D + 0.55L + 1.1E_h$	Negative Positive	37.1 -814.3 194.2
5	$0.77D + 1.1E_h$	Negative Positive	251.4 -600.0 87.1

The required reinforcement for the beam on the fourth floor level is shown in Table 5-15. The provided areas of steel are within the limits specified in 1921.3.2.1. Also given in Table 5-15 are the flexural moment strengths (ϕM_n) at each section. Note that at the supports, a sufficient amount of positive reinforcement must be provided in order to satisfy the provision of 1921.3.2.2: the positive moment strength at a joint face must be at least equal to 50% of the negative moment strength provided at that joint. At the supports, the positive moment strength must be at least equal to 835.1/2 = 417.6 ft-kips. Providing 6-No. 8 bars (ϕM_n = 432.1 ft-kips) as positive reinforcement satisfies this requirement. Additionally, neither the negative nor the positive moment strength at any section along the length of the member shall be less than 25% of the maximum moment strength provided at the face of either joint (1921.3.2.2). In this case, 25% of the maximum moment strength is equal to 835.1/4 = 208.8 ft-kips. Providing at least 2-No. 10 bars (ϕM_n = 240.8 ft-kips) or 3-No. 8 bars (ϕM_n = 225.3 ft-kips) at any section will satisfy this requirement, as well as the minimum reinforcement requirement of 1921.3.2.1. This also automatically satisfies the requirement that 2 bars be continuous at both the top and bottom of any section (1921.3.2.1).

When reinforcing bars extend through a joint, the column dimension must be at least 20 times the diameter of the largest longitudinal bar for normal weight concrete (1921.5.1.4). In this case, the minimum required column dimension is 20 × 1.27 = 25.4 in. which is less than each of the two 36 in. wide columns that are provided.

Table 5-15 Summary of Required Reinforcement for Beam C2-D2 at Floor Level 4

	M_u (ft-kips)	A_s* (in.^{2})	Reinforcement	ϕM_n (ft-kips)
Negative	-814.3	9.84	8-No. 10	835.1
Positive	220.4	2.32	6-No. 8	432.1

*Max. $A_s = 0.025 \times 24 \times 22 = 13.20$ in.^{2} (1921.3.2.1)

Min. $A_s = 3\sqrt{4{,}000} \times 24 \times 22/60{,}000 = 1.67$ in.^{2}

$= 200 \times 24 \times 22/60{,}000 = 1.76$ in.^{2} (governs)

Design for Shear.

The beams must be designed for shear forces in accordance with 1921.3.4. These shear forces are determined assuming that moments of opposite sign corresponding to the probable moment strength (M_{pr}) act at the joint faces and that the member is loaded with the tributary gravity load along its span. It is important to note that sidesway to the right and to the left must both be considered when calculating the maximum design shear forces.

The probable moment strength (M_{pr}) is determined for a section using the stress in the tensile reinforcement equal to $1.25f_y$ and a strength reduction factor (ϕ) equal to 1.0. The following equation can be used to compute M_{pr} :

$$M_{pr} = A_s(1.25f_y)\left(d - \frac{a}{2}\right)$$

where $a = \dfrac{A_s(1.25f_y)}{0.85f_c' b}$

For example, for sidesway to the right, the right joint must be subjected to the negative moment M_{pr} which is determined as follows:

For 8-No. 10 top bars, $A_s = 8 \times 1.27 = 10.16$ in.^{2}

$$a = \frac{A_s(1.25f_y)}{0.85f_c' b} = \frac{10.16 \times 1.25 \times 60}{0.85 \times 4 \times 24} = 9.34 \text{ in.}$$

$$M_{pr} = A_s(1.25f_y)\left(d - \frac{a}{2}\right) = 10.16 \times 1.25 \times 60 \times \left(22 - \frac{9.34}{2}\right) = 13{,}206 \text{ in.-kips} = 1{,}100.5 \text{ ft-kips}$$

Similarly, for the left joint, the positive moment M_{pr} based on the 6-No. 8 bottom bars is equal to 587.2 ft-kips. The probable moment strengths for sidesway to the left can be obtained in a similar fashion.

The gravity load on the beam is the same as the gravity load on beam A3-B3, i.e., $w = 5{,}744$ plf.

Figure 5-5 shows beam C2-D2 and the probable moment strengths (M_{pr}) at the joint faces for sidesway to the right and to the left. This figure also shows the tributary gravity loads on the beam, as well as the combined shear forces at the faces of the columns.

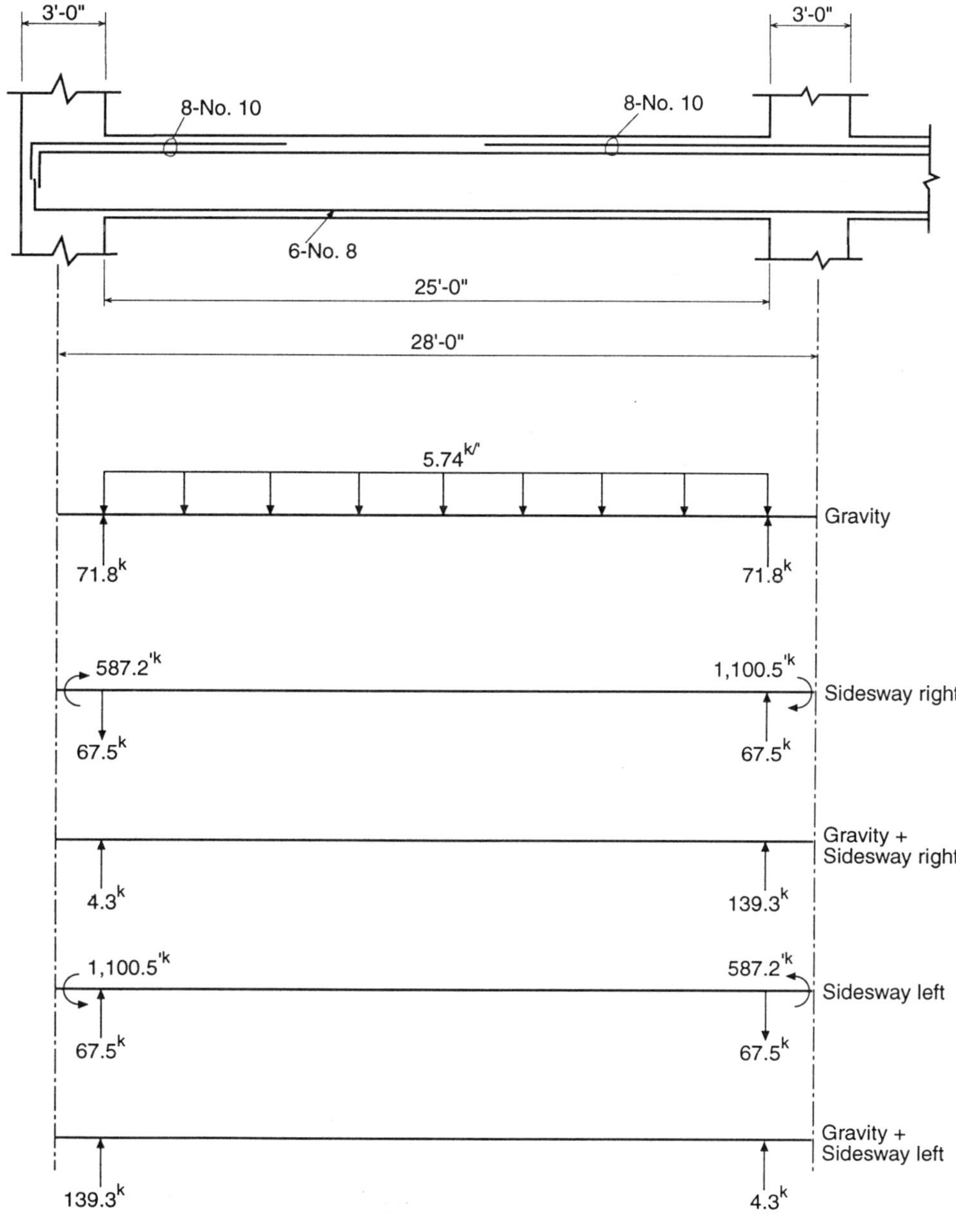

Figure 5-5 Design Shear Forces for Beam C2-D2

In general, the shear strength is assumed to be provided by both concrete (V_c) and reinforcing steel in the form of hoops or stirrups (V_s). However, the shear strength of the concrete (V_c) is to be taken as zero when the earthquake-induced shear force calculated in accordance with 1921.3.4.1 is greater than or equal to 50% of the total shear force and the factored axial compressive force is less than $A_g f'_c /20$ where A_g is the gross area of the beam (1921.3.4.2). In this particular example, the beam carries negligible axial forces, and the maximum earthquake-induced shear force, which is equal to (1,100.5 + 587.2)/25 = 67.5 kips, is less than one-half the total design shear force which is equal to 0.5 × 139.3 = 69.7 kips. However, at sections located more than approximately 9 in. from the face of the support, the earthquake-induced shear force is greater than one-half of the total design shear. Thus, V_c is set equal to zero at every section located within $2h$ = 49 in. = 4.1 ft from the face of the support. The maximum shear force V_s is (1911.1):

$$\phi V_s = V_u - \phi V_c$$

$$V_s = \frac{V_u}{\phi} - 0 = \frac{139.3}{0.85} = 163.9 \text{ kips}$$

Shear strength contributed by shear reinforcement shall not exceed $(V_s)_{max}$ (1911.5.6.8):

$$(V_s)_{max} = 8\sqrt{f_c'}\, b_w d = 8\sqrt{4{,}000} \times 24 \times 22/1{,}000 = 267.2 \text{ kips} > 163.9 \text{ kips} \quad \text{O.K.}$$

Required spacing of No. 4 closed stirrups (hoops) for a factored shear force of 163.9 kips is (1911.5.6.2):

$$s = \frac{A_v f_y d}{V_s} = \frac{(5 \times 0.2) \times 60 \times 22}{163.9} = 8.1 \text{ in.}$$

Note that 5 legs are required in order to conform with the provisions of 1921.3.3.3 for lateral support of the longitudinal bars.

Maximum allowable hoop spacing (s_{max}) within a distance of $2h = 2 \times 24.5 = 49$ in. (1921.3.3.1) from the face of the support is the smallest of the following (1921.3.3.2):

$$s_{\max} = \frac{d}{4} = \frac{22}{4} = 5.5 \text{ in. (governs)}$$

$$= 8 \times (\text{diameter of smallest longitudinal bar}) = 8 \times 1.0 = 8.0 \text{ in.}$$

$$= 24 \times (\text{diameter of hoop bar}) = 24 \times 0.50 = 12.0 \text{ in.}$$

$$= 12 \text{ in.}$$

Therefore, hoops must be spaced at 5 in. on center with the first one located at 2 in. from the face of the support. Eleven hoops are to be placed at this spacing. Where hoops are no longer required, stirrups with seismic hooks at both ends may be used (1921.3.3.4). At a distance of 52 in. from the face of the support, V_u = 114.4 kips. Also, with V_c = 66.8 kips, the spacing required for No. 4 stirrups with 2 legs is 7.8 in., which is less than the maximum allowable spacing of $d/2 = 11.0$ in (1921.3.3.4). A 7 in. spacing, starting at 52 in. from the face of the support, will be sufficient for the remaining portion of the beam.

Reinforcing Bar Cutoff Points.

The negative reinforcement at the supports is 8-No. 10 bars. The location where five of the eight bars can be terminated will be determined.

The loading used to find the cutoff point of the 5-No. 10 bars is 0.9 times the dead load in combination with the probable flexural strengths (M_{pr}) at the ends of the member; this combination will produce the longest bar lengths. The design flexural strength (ϕM_n) provided by 3-No. 10 bars is 353.2 ft-kips. Therefore, the 5 reinforcing bars can be terminated after the required moment strength (M_u) has been reduced to 353.2 ft-kips.

With ϕ = 1.0 and f_s = 75 ksi, M_{pr} = 1,100.5 ft-kips at one support and 587.2 ft-kips at the other. The load is equal to 0.9 × {[(109 + 25) × 28] + [(24 × 24.5/144) × 145]}/1,000 = 3.91 kips/ft. The distance from the face of the support to where the moment equals 353.2 ft-kips is obtained by summing moments about section *a-a* in Fig. 5-6:

$$3.91x\left(\frac{x}{2}\right) - 116.4x + 1{,}100.5 = 353.2$$

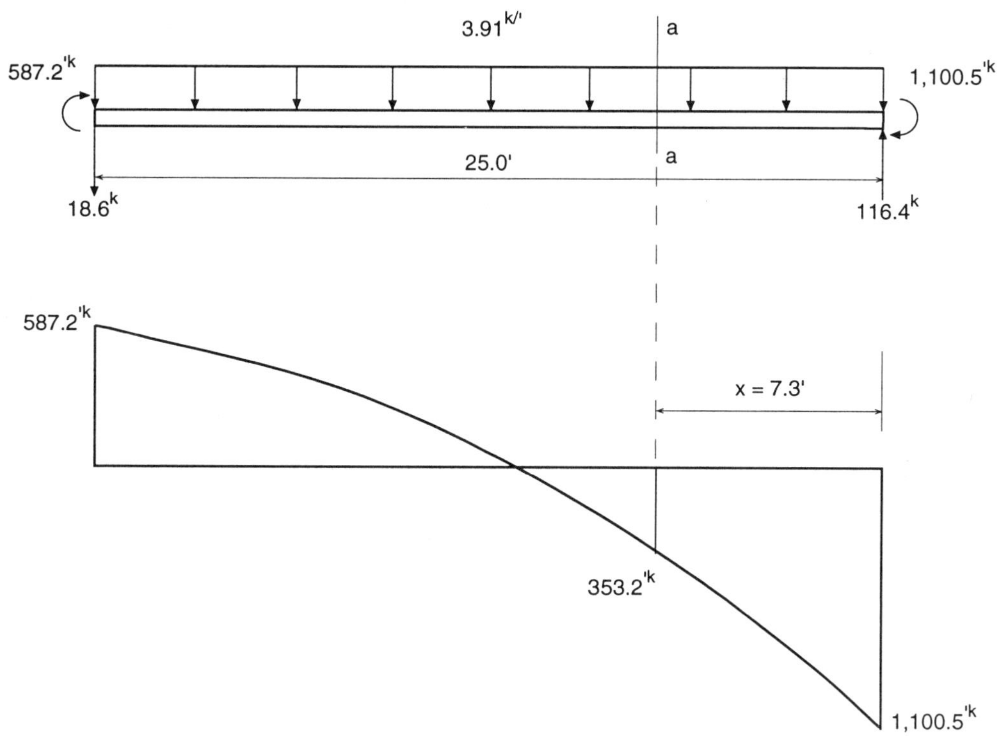

Figure 5-6 Moment Diagram for Cutoff Location of Negative Bars at Supports of Beam C2-D2

Solving for *x* gives a distance of 7.3 ft. The 5-No. 10 bars must extend a distance d = 22 in. or $12d_b$ = 12 × 1.27 = 15.2 in. beyond the distance *x* (1912.10.3). Thus, from the face of the support, the total bar length must be at least equal to 7.3 + (22/12) = 9.1 ft. Also, the bars must extend a full development length ℓ_d beyond the face of the support (1912.10.4). The development length can be determined from the provisions of 1912.2.3:

$$\frac{\ell_d}{d_b} = \frac{3}{40}\frac{f_y}{\sqrt{f_c'}}\frac{\alpha\beta\gamma\lambda}{\left(\dfrac{c + K_{tr}}{d_b}\right)}$$

where α = reinforcement location factor = 1.3 (top bar)

β = coating factor = 1.0 (uncoated reinforcement)

γ = reinforcement size factor = 1.0 (No. 10 bar)

λ = lightweight aggregate concrete factor = 1.0 (normal weight concrete)

$$c = \text{ spacing or cover dimension } = \begin{cases} 1.5 + 0.5 + \dfrac{1.27}{2} = 2.64 \text{ in.} \\ \dfrac{24 - 2(1.5 + 0.5) - 1.27}{2 \times 7} = 1.34 \text{ in. (governs)} \end{cases}$$

K_{tr} = transverse reinforcement index = 0 (conservative)

$$\frac{c + K_{tr}}{d_b} = \frac{1.34 + 0}{1.27} = 1.06$$

$$\frac{\ell_d}{d_b} = \frac{3}{40} \times \frac{60,000}{\sqrt{4,000}} \times \frac{1.3 \times 1.0 \times 1.0 \times 1.0}{1.06} = 87.3$$

$$\ell_d = 87.3 \times 1.27 = 110.8 \text{ in.} = 9.2 \text{ ft} > 9.1 \text{ ft}$$

Thus, the total required length of the 5-No. 8 bars must be at least 9.2 ft beyond the face of the support. It is important to note that flexural reinforcement shall not be terminated in a tension zone unless one or more of the conditions of 1912.10.5 are satisfied. In this case, the point of inflection is approximately 11.8 ft from the face of the support which is greater than 9.2 ft. Therefore, the 5-No. 10 bars can not be terminated here unless one of the conditions of 1912.10.5 is satisfied. Using 1912.10.5.1, check if the factored shear force (V_u) at the cutoff point does not exceed two-thirds of that permitted ($2\phi V_n/3$). For No. 4 stirrups at 7 in. on center that are provided in this region of the beam:

$$\phi V_n = \phi(V_s + V_c) = 0.85 \times \left(\frac{0.4 \times 60 \times 22}{7} + 66.8 \right) = 120.9 \text{ kips}$$

$$\frac{2}{3} \phi V_n = 80.6 \text{ kips} > V_u = 80.4 \text{ kips at 9.2 ft from face of support}$$

Since V_u is less than $2\phi V_n/3$, the cutoff point for the 5-No. 10 bars can be 9.2 ft beyond the face of the support.

Figure 5-7 shows the reinforcement details for beam C2-D2.

5.6.4 Column Design

This section outlines the design of column A3 between the fifth and sixth floor levels. Since this column is not part of the LFR system, the provisions of 1921.7 must be satisfied. A summary of the design forces is given in Tables 5-16. Note that orthogonal load effects on the column need not be considered, since this column is not part of two or more intersecting lateral force resisting systems (1633.1).

Based on the load combinations in Table 5-16, a 24 × 24 in. column with 12-No. 10 bars (ρ_g = 2.65%) is adequate. Figure 5-8, obtained from *PCACOL* (Ref. 5.1), shows the interaction diagram for this column.

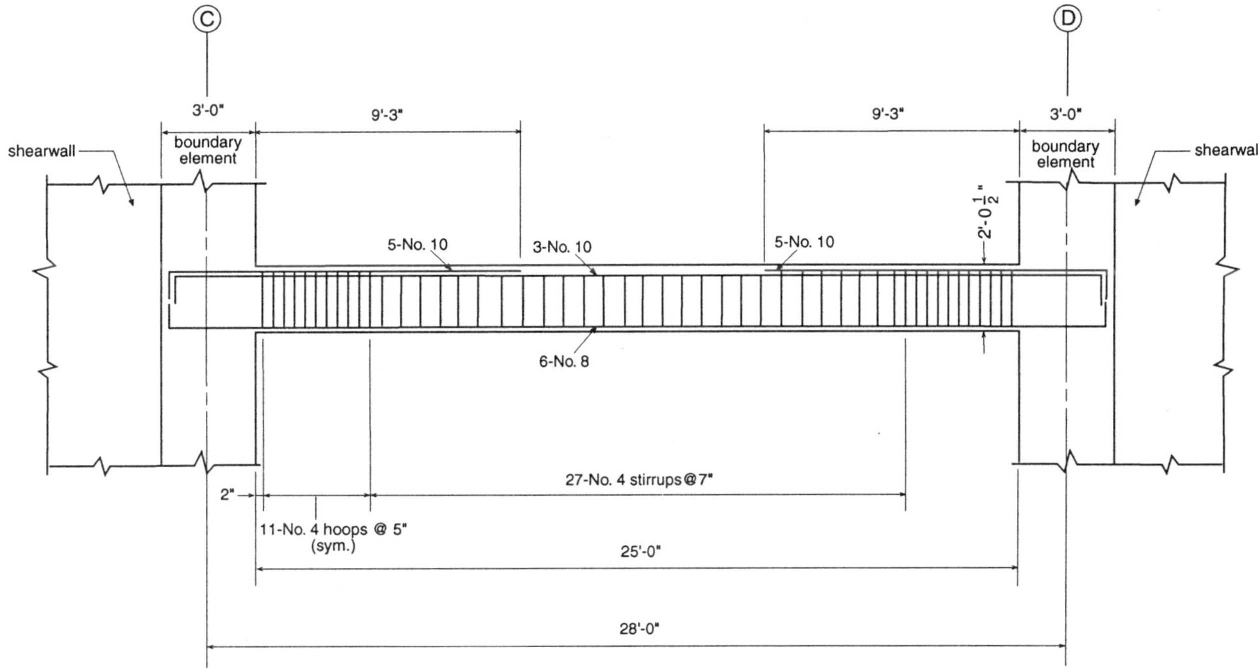

Figure 5-7 Reinforcement Details for Beam C2-D2

Table 5-16 Summary of Factored Axial Loads and Bending Moments for Column A3 between the 5th and 6th Floor Levels

No.	Load Case	Axial Load (kips)	Bending Moment (ft-kips)
	Dead (*D*)	410	-88.3*
	Live (*L*)	40	-28.5*
	Earthquake (*E*)	±246	±399.5
No.	Load Combination		
1	$1.4D + 1.4L + E$	384	236.0
		876	-563.0
2	$0.9D + E$	123	320.0
		615	-479.0

**beam A3-B3 exterior negative bending moment/2 (see Table 5-7)*

The governing shear strength and detailing requirements depend on the magnitude of the factored gravity axial force. In this case, the maximum factored axial load due the gravity loads is $1.4 \times (410 + 40) = 630$ kips which is greater than $A_g f'_c /10 = 24^2 \times 4/10 = 230$ kips. Also, the 630 kip axial load must be compared to $0.3P_o$ where P_o is the nominal axial load capacity of the column at zero eccentricity:

$$0.3P_o = 0.3 \times [0.85 f'_c(A_g - A_{st}) + f_y A_{st}] = 0.3 \times \{[0.85 \times 4 \times (24^2 - 15.24)] + (60 \times 15.24)\} = 846 \text{ kips}$$

Since the maximum factored gravity axial force is less than $0.3P_o$, the provisions of 1921.7.2.2 must be satisfied. Note that the provided reinforcement ratio is within the allowable range of 1% and 6% (1921.4.3.1).

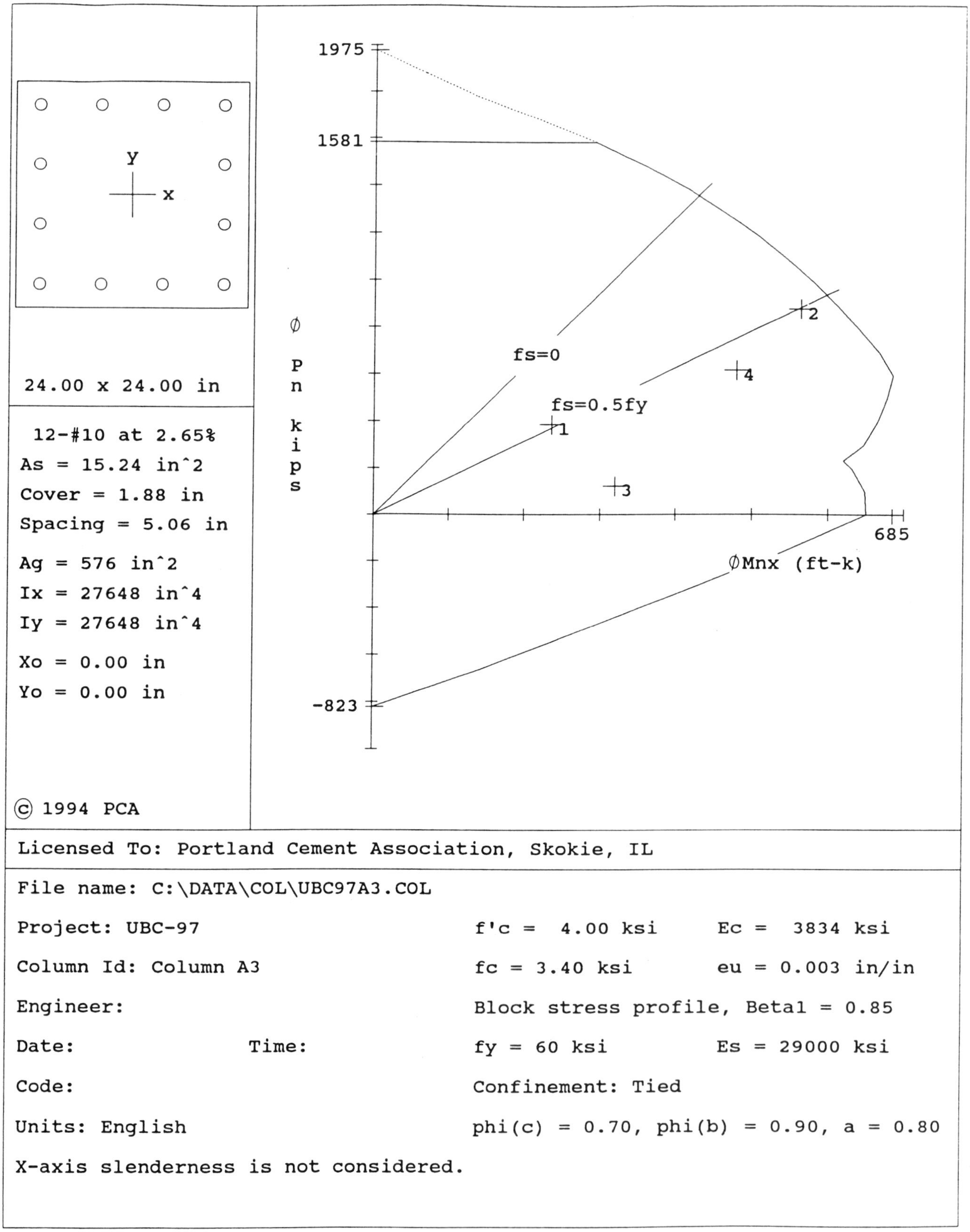

Figure 5-8 Design Strength Interaction Diagram for Column A3 between the 5th and 6th Floor Levels

Confinement Reinforcement.

The longitudinal spacing of ties for the full column height shall not exceed s_o (1921.7.2.2):

$$s_o = 6 \times \text{(diameter of smallest longitudinal bar)} = 6 \times 1.27 = 7.6 \text{ in.}$$
$$= 16 \times \text{(diameter of tie bar)} = 16 \times 0.5 = 8 \text{ in.}$$
$$= 0.5 \times \text{(least cross-sectional dimension)} = 0.5 \times 24 = 12 \text{ in.}$$
$$= 6 \text{ in. (governs)}$$

Thus, tentatively select No. 4 ties with one crosstie spaced at 6 in. for the full column height.

Transverse Reinforcement for Shear.

According to 1921.7.2.2, the column must be designed for shear forces associated with the development of the nominal moment strengths (M_n) of the member at each end of the clear span. It is important to note that the calculated shear force must not be taken less than the factored shear force obtained from analysis.

Conservatively assume that the nominal moment strength (M_n) is equal to the moment at the balanced point which is approximately 978 ft-kips (see Fig. 5-9). Thus, the shear force at the column ends produced by these bending moments is equal to (2 × 978)/[13 – (24.5/12)] = 178.5 kips.

The shear force caused by gravity loads is determined by adding the bending moments at each end of the column and dividing by the clear column height. Therefore, the unfactored shear forces due to the dead and live loads are (see Table 5-16):

$$V_D = \frac{M_D}{\ell_n} = \frac{2 \times 88.3}{\left(13 - \frac{24.5}{12}\right)} = 16.1 \text{ kips}$$

$$V_L = \frac{M_L}{\ell_n} = \frac{2 \times 28.5}{\left(13 - \frac{24.5}{12}\right)} = 5.2 \text{ kips}$$

From Table 5-5, the shear force from deformation compatibility (V_E) is 55.9 kips. Based on the load combination of 1921.7.2, the factored shear force (V_u) is:

$$V_u = 1.4V_D + 1.4V_L + V_E = 1.4 \times (16.1 + 5.2) + 55.9 = 85.7 \text{ kips} < 178.5 \text{ kips}$$

For members subjected to axial compression, the shear strength provided by the concrete (V_c) can be determined from Eq. (11-4):

$$V_c = 2\left(1 + \frac{N_u}{2{,}000A_g}\right)\sqrt{f_c'}\, b_w d$$

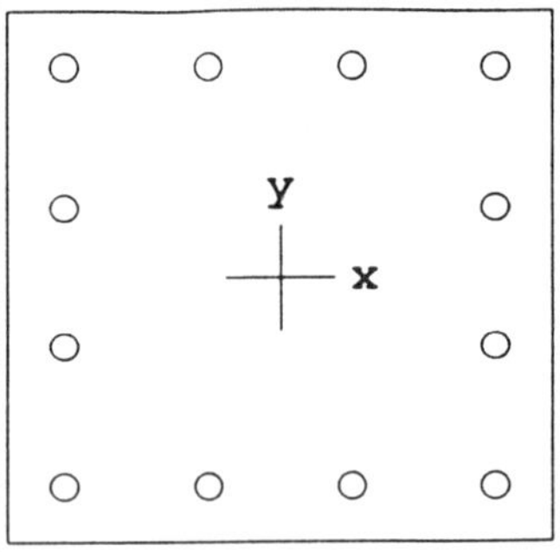

24.00 x 24.00 in

12-#10 at 2.65%
As = 15.24 in^2
Cover = 1.88 in
Spacing = 5.06 in

Ag = 576 in^2
Ix = 27648 in^4
Iy = 27648 in^4

Xo = 0.00 in
Yo = 0.00 in

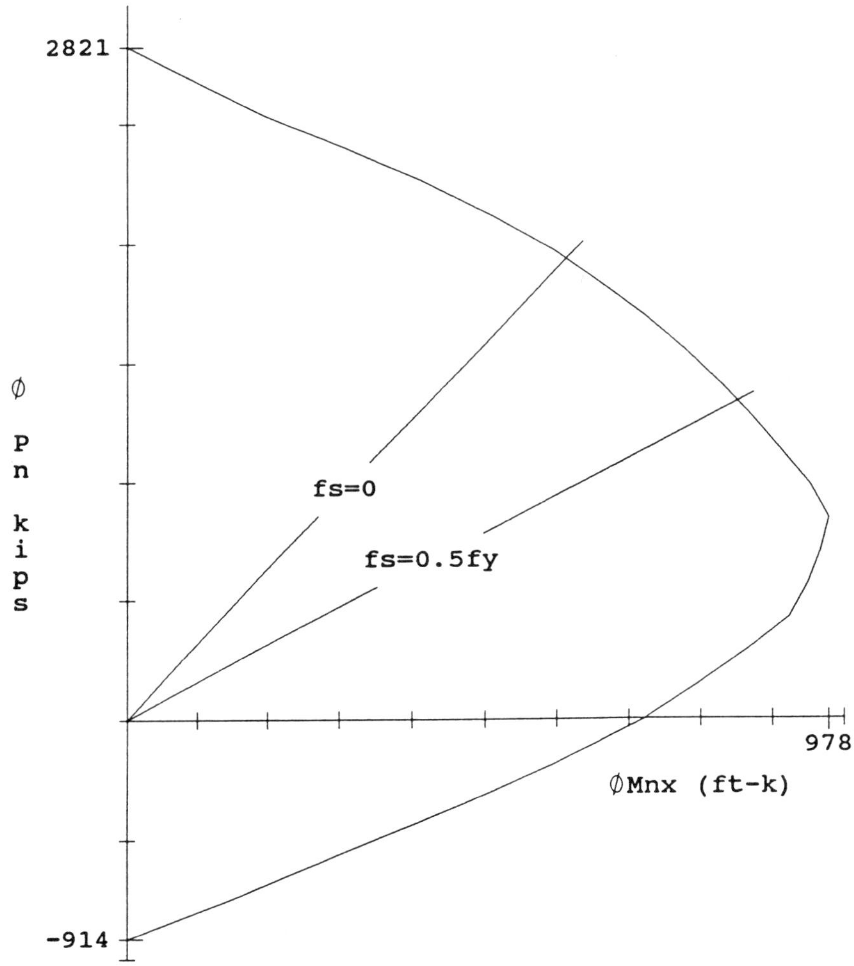

File name: C:\DATA\COL\UBC97A3.COL

Project: UBC-97		f'c = 4.00 ksi	Ec = 3834 ksi
Column Id: Column A3		fc = 3.40 ksi	eu = 0.003 in/in
Engineer:		Block stress profile, Beta1 = 0.85	
Date:	Time:	fy = 60 ksi	Es = 29000 ksi
Code:		Confinement: Other	
Units: English		phi(c) = 1.00, phi(b) = 1.00, a = 1.00	

X-axis slenderness is not considered.

Figure 5-9 Design Strength Interaction Diagram for Column A3 with $\phi = 1.0$

Using the minimum factored axial load (N_u) on the column (see Table 5-16),

$$V_c = 2\left(1 + \frac{123{,}000}{2{,}000 \times 24^2}\right)\sqrt{4{,}000} \times 24 \times 19.3/1{,}000 = 64.9 \text{ kips}$$

The shear strength (V_s) provided by the No. 4 ties with one crosstie spaced at 6 in. is (1911.5.6.2, 1911.5.6.8):

$$V_s = \frac{A_v f_y d}{s} = \frac{(3 \times 0.2) \times 60 \times 19.3}{6} = 115.8 \text{ kips} < 8\sqrt{f_c'}\, b_w d = 234.4 \text{ kips}$$
$$< 4\sqrt{f_c'}\, b_w d = 117.2 \text{ kips}$$

The design shear strength (ϕV_n) of the column is:

$$\phi V_n = \phi(V_c + V_s) = 0.85(64.9 + 115.8) = 153.6 \text{ kips} < 178.5 \text{ kips} \quad \text{N.G.}$$

Thus, the No. 4 ties with one crosstie at 6 in. are not adequate for shear strength requirements. Providing No. 4 ties with two crossties at 6 in. on center results in a design shear strength of ϕV_n = 186.4 kips which is greater than the required shear strength. However, in this case, V_s is equal to 154.4 kips which is greater than $4\sqrt{f_c'} b_w d$ = 117.2 kips. Therefore, the maximum spacing of the shear reinforcement is $d/4$ = 19.3/4 = 4.8 in. (1911.5.4.3). Considering the conservative assumptions that were made above, use No. 4 ties with two crossties spaced at 5 in. on center for the full column height. Note that the provisions of 1921.4.4.3 are satisfied with this tie arrangement.

Lap Splice Length for Column Vertical Bars.

The lap splice length of the No. 10 bars in column A3 can be determined according to 1912.2.3:

$$\frac{\ell_d}{d_b} = \frac{3}{40}\frac{f_y}{\sqrt{f_c'}}\frac{\alpha\beta\gamma\lambda}{\left(\frac{c + K_{tr}}{d_b}\right)}$$

where α = reinforcement location factor = 1.0 (no top bar effect)

β = coating factor = 1.0 (uncoated reinforcement)

γ = reinforcement size factor = 1.0 (No. 7 or larger bar size)

λ = lightweight aggregate concrete factor = 1.0 (normal weight concrete)

$$c = \text{spacing or cover dimension} = \begin{cases} 1.5 + 0.5 + \dfrac{1.27}{2} = 2.6 \text{ in. (governs)} \\ \dfrac{24 - 2(1.5 + 0.5) - 1.27}{2 \times 3} = 3.1 \text{ in.} \end{cases}$$

$$K_{tr} = \text{transverse reinforcement index} = \frac{A_{tr} f_{yt}}{1{,}500 sn} = \frac{4 \times 0.2 \times 60{,}000}{1{,}500 \times 5 \times 4} = 1.6$$

$$\frac{c + K_{tr}}{d_b} = \frac{2.6 + 1.6}{1.27} = 3.3 > 2.5, \text{ use } 2.5$$

$$\frac{\ell_d}{d_b} = \frac{3}{40} \times \frac{60,000}{\sqrt{4,000}} \times \frac{1.0 \times 1.0 \times 1.0 \times 1.0}{2.5} = 28.5$$

$$\ell_d = 28.5 \times 1.27 = 36.2 \text{ in.} = 3.1 \text{ ft}$$

According to 1921.7.2.2, the lap splices must conform to the provisions of 1921.4.3.2, i.e., the splices must be located within the center half of the column length and must be proportioned as tension splices. From Fig. 5-8 it can be seen that a Class B splice is required since the bar stress is greater than $0.5f_y$ in tension for at least one of the load combinations (1912.17.2.3). Therefore, the Class B splice length is equal to 1.3 × 3.1 = 4.0 ft. Reinforcement details for column A3 are shown in Fig. 5-10.

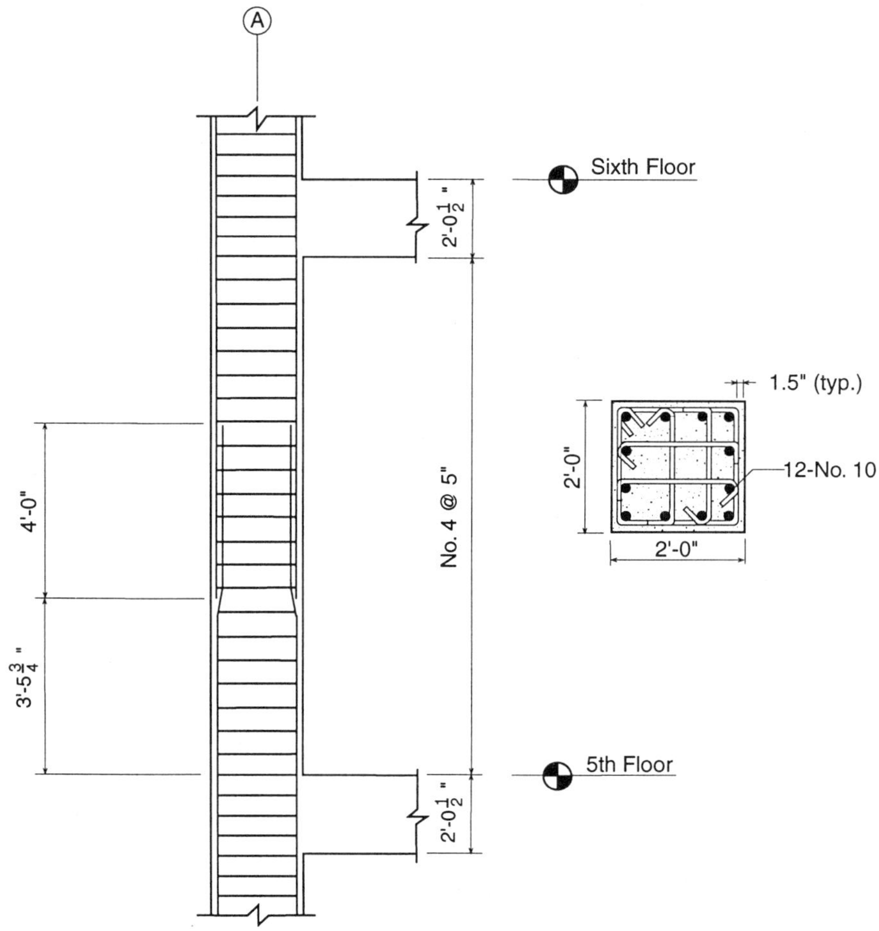

Figure 5-10 Reinforcement Details for Column A3

5.6.5 Shearwall Design

The design procedure for shearwalls can be summarized in three steps:

1. Design the shearwall for flexure and axial load by considering the entire cross-section of the wall to act as a short column (1921.6.6.1); this will yield the required flexural reinforcement. Check the shear resistance of the wall (1921.6.5).

2. Provide the boundary zone detail requirements of 1921.6.6.6 at each end of the wall unless $P_u \leq 0.10 A_g f'_c$ (for geometrically symmetrical walls) and either $M_u / V_u \ell_w \leq 1.0$ or $V_u \leq 3A_{cv}\sqrt{f'_c}$ and $M_u / V_u \ell_w \leq 3.0$, where ℓ_w is the length of the entire wall in the direction of the shear force and A_{cv} is the net area of the concrete section bounded by the web thickness and the length of the section in the direction of the shear force (1921.6.6.4). In any case, shearwalls with $P_u > 0.35 P_o$ are not permitted to resist earthquake-induced forces, where P_o is the nominal axial load strength of the wall at zero eccentricity (1921.6.6.3).

3. For shearwalls where boundary elements with specially detailed reinforcement have to be provided, determine the length of the boundary zone by using one of two methods. In the conservative method, the boundary zones are provided over a distance of $0.25\ell_w$ to $0.15\ell_w$ at each end of the wall (1921.6.6.4). Alternatively, the boundary zone length may be based on the compressive strain levels at the edges of the wall when the wall is subjected to displacements resulting from the ground motions specified in 1629.2 using cracked section properties of the wall (1921.6.6.5). Boundary zone detail requirements of 1921.6.6.6 must be provided wherever the compressive strain exceeds 0.003.

Combined Load Effects.

A summary of the design forces at the base of shearwall D2-E2 is given in Table 5-17 (see Tables 5-3 and 5-10).

Table 5-17 Summary of Factored Axial Loads, Bending Moments, and Shear Forces at the Base of Shearwall D2-E2

Load Case		Axial Load (kips)	Bending Moment (ft-kips)	Shear Force (kips)
	Dead (D)	3,238	0	0
	Live (L)	298	0	0
	Earthquake (E_h)	±286	±111,052	±1,256
No.	Load Combination			
1	$1.4D + 1.7L$	5,040	0	0
4	$1.54D + 0.55L + 1.1E_h$	4,836	122,157	1,382
		5,465	-122,157	-1,382
5	$0.77D + 1.1E_h$	2,179	122,157	1,382
		2,808	-122,157	-1,382

Shear Design.

At least two curtains of reinforcement must be used in the wall between the boundary elements if the in-plane shear force (V_u) exceeds $2A_{cv}\sqrt{f'_c}$ where A_{cv} is the net area bounded by the web thickness and the length of the section in the direction of analysis (1921.6.2.2):

$$V_u = 1{,}382 \text{ kips} > 2A_{cv}\sqrt{f'_c} = 2 \times (12 \times 31 \times 12) \times \sqrt{6{,}000}/1{,}000 = 692 \text{ kips}$$

Thus, two curtains of reinforcement are required in the web. Note that the reinforcement ratio along the longitudinal and transverse directions must not be less than 0.0025, and that the spacing of the bars must not exceed 18 in. (1921.6.2.1).

The upper limit on the shear strength of a shearwall is $\phi 8 A_{cv}\sqrt{f_c'}$ which is equal to 0.6 × 4 × 692 = 1,661 kips (1921.6.5.6). Typically, the shear strength reduction factor is 0.85. However, this factor must be taken as 0.6 for any structural member if its nominal shear strength is less than the shear corresponding to the development of the nominal flexural strength of the member (1909.3.4.1). In this case, the more conservative value of 0.6 was used in the design of the wall; it is clear that the factored shear force is less than the limiting value.

Since the ratio of the height of the wall (h_w) to the length of the wall (ℓ_w) is equal to 133/31 = 4.3, the design shear strength of the wall (ϕV_n) is given by the following (1921.6.5.3):

$$\phi V_n = \phi A_{cv}(2\sqrt{f_c'} + \rho_n f_y)$$

where ρ_n is equal to the ratio of distributed shear reinforcement on a plane perpendicular to the plane of A_{cv}.

Using 2-No. 5 bars at 8 in. on center gives the shear strength of the wall as follows:

$$\rho_n = \frac{2 \times 0.31}{12 \times 8} = 0.0065$$

$$\phi V_n = 0.6 \times (12 \times 31 \times 12) \times \left[2\sqrt{6{,}000} + (0.0065 \times 60{,}000)\right] / 1{,}000 = 1{,}460 \text{ kips} > 1{,}382 \text{ kips O.K.}$$

The vertical reinforcement ratio (ρ_v) shall not be less than the horizontal reinforcement ratio (ρ_n) when the ratio h_w / ℓ_w is less than 2.0 (1921.6.5.5). Since h_w / ℓ_w is equal to 4.3, the minimum reinforcement ratio will be used. The required area of vertical steel in the web is equal to 0.0025 × 12 × 12 = 0.36 in.^{2}/ft. Using 2 layers of No. 5 bars, the required spacing is (0.62 × 12/0.36) = 20.7 in. which is greater than the maximum allowable spacing of 18 in. Thus, use No. 5 vertical bars on each face at 18 in. on center.

Design of Boundary Zone.

Determine if boundary zone details are required. From Table 5-17, the maximum factored axial load (P_u) is equal to 5,465 kips. Check the criterion of 1921.6.6.4:

$$0.10 A_g f_c' = 0.10 \times [(12 \times 25 \times 12) + (2 \times 36^2)] \times 6.0 = 3{,}715 \text{ kips} < P_u = 5{,}465 \text{ kips}$$

Therefore, boundary zone details of 1921.6.6.6 are required. Assume that the 36 × 36 in. boundary elements are reinforced with 36-No. 14 bars. It was determined above that 2-No. 5 bars at 18 in. spacing is required in the web. Based on this reinforcement, the nominal axial load capacity of the wall at zero eccentricity (P_o) is:

$$P_o = 0.85 f_c'(A_g - A_{st}) + f_y A_{st}$$

$$= [0.85 \times 6 \times (6{,}192 - 171.92)] + (60 \times 171.92) = 41{,}018 \text{ kips}$$

Since P_u is equal to 5,465 kips which is $0.13P_o$, conservatively provide boundary zones at each end a distance of $0.15\,\ell_w$ = 0.15 × 31 = 4.7 ft (1921.6.6.4).

Alternatively, determine the length of the boundary zone using the provisions of 1921.6.6.5. Before proceeding with the strain compatibility analysis, determine if the wall can be considered to contribute to the strength of the structure for resisting earthquake-induced forces. Since P_u is equal to 5,465 kips which is less than $0.35P_o$ = 14,356 kips, the wall can be used to resist its portion of the seismic loads.

When performing the strain compatibility analysis for the wall, it is important to note that the compressive strain in the concrete is not allowed to be larger than 0.015 (see Eq. (21-8)).

For shearwalls in which the flexural limit state is governed by yielding at the base of the wall, the total curvature demand (ϕ_t) can be determined from Eq. (21-9):

$$\phi_t = \frac{\Delta_i}{(h_w - \ell_p / 2)\ell_p} + \phi_y$$

where Δ_i = inelastic deflection at the top of the wall = Δ_t - Δ_y

Δ_t = total deflection at the top of the wall equal to Δ_M, using cracked section properties, or may be taken as $2\Delta_M$, using gross section properties

Δ_M = maximum inelastic response displacement = $0.7R\Delta_S$

Δ_S = design level response displacement

Δ_y = displacement at top of wall corresponding to yielding of the tension reinforcement at the critical section, or may be taken as $(M'_n / M_E)\Delta_E$, where M_E is equal to the unfactored moment at the critical section when the top of the wall is displaced Δ_E and M'_n is the nominal flexural strength of the critical section at an axial load $P'_u = 1.2D + 0.5L + E$

Δ_E = elastic design displacement at the top of the wall using gross section properties and code-specified seismic forces

ℓ_p = height of the plastic hinge above the critical section, which may be taken as $0.5\ell_w$

ϕ_y = yield curvature, which may be estimated as $0.003/\ell_w$

Note that if ϕ_t is less than or equal to $0.003/c'_u$, where c'_u is the neutral axis depth at P'_u and M'_n, boundary zone details are not required. If ϕ_t exceeds $0.003/c'_u$, the compressive strains may be assumed to vary linearly over the depth c'_u and have a maximum value equal to the product of c'_u and ϕ_t.

The first step in the analysis is to determine the location of the neutral axis c'_u. Using Table 5-17, the load P'_u is equal to (1.2 × 3,238) + (0.5 × 298) + 286 = 4,321 kips. The nominal strength (M'_n) corresponding to P'_u is 206,307 ft-kips (see Fig. 5-11).

In lieu of hand computations, *PCACOL* (Ref. 5.1) was used to obtain the neutral axis depth c'_u corresponding to P'_u and M'_n. For 36 × 36 in. boundary elements reinforced with 36-No. 14 bars and a 12 in. web reinforced with 2-No. 5 bars at 18 in., c'_u is equal to 46.7 in.

Figure 5-12 shows the interaction diagram with the load combinations that are contained in Table 5-17. As can be seen from the figure, the wall is adequate to carry the factored loads.

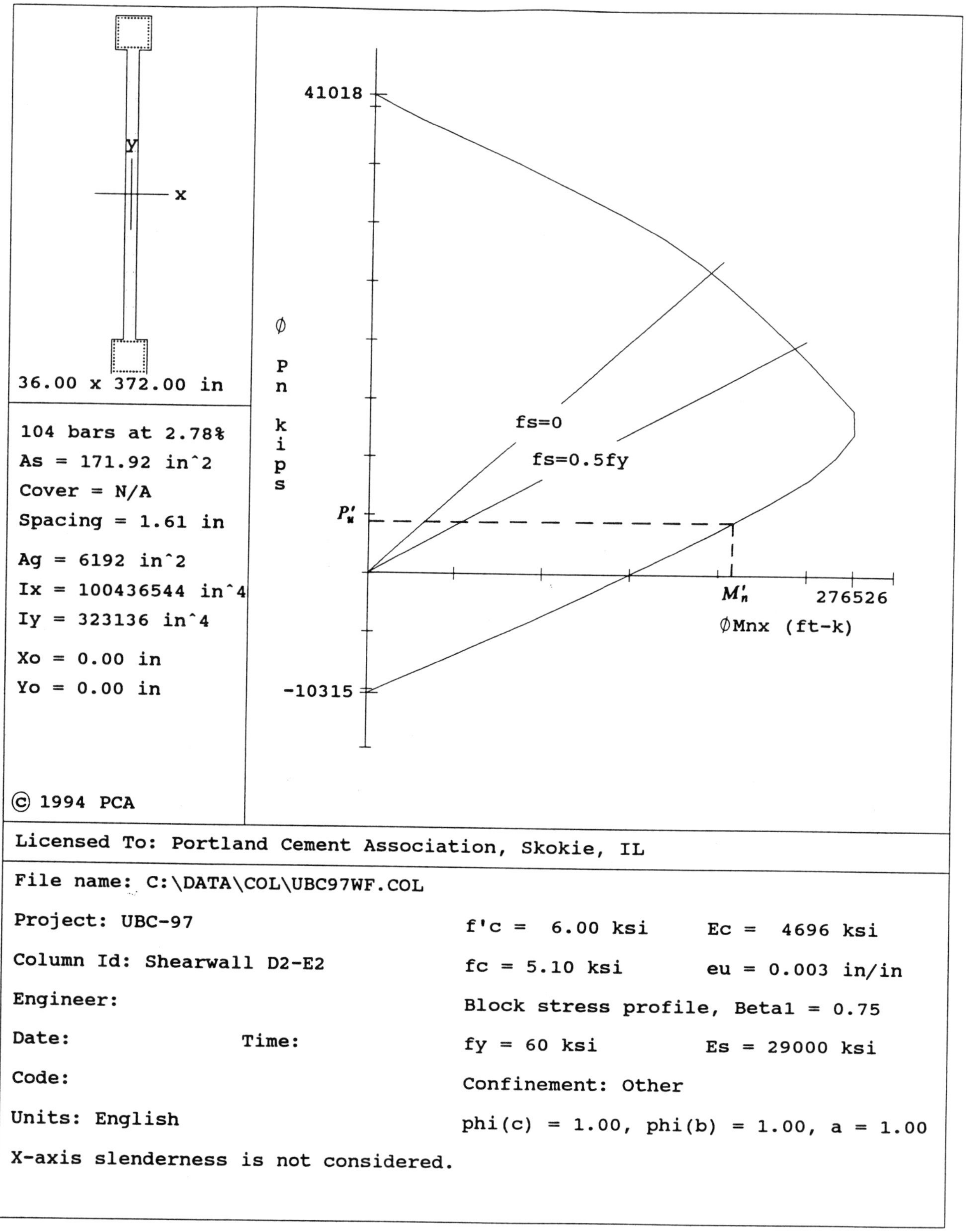

Licensed To: Portland Cement Association, Skokie, IL

File name: C:\DATA\COL\UBC97WF.COL

Project: UBC-97	f'c = 6.00 ksi Ec = 4696 ksi
Column Id: Shearwall D2-E2	fc = 5.10 ksi eu = 0.003 in/in
Engineer:	Block stress profile, Beta1 = 0.75
Date: Time:	fy = 60 ksi Es = 29000 ksi
Code:	Confinement: Other
Units: English	phi(c) = 1.00, phi(b) = 1.00, a = 1.00

X-axis slenderness is not considered.

Figure 5-11 $P_n - M_n$ Diagram for Shearwall D2-E2

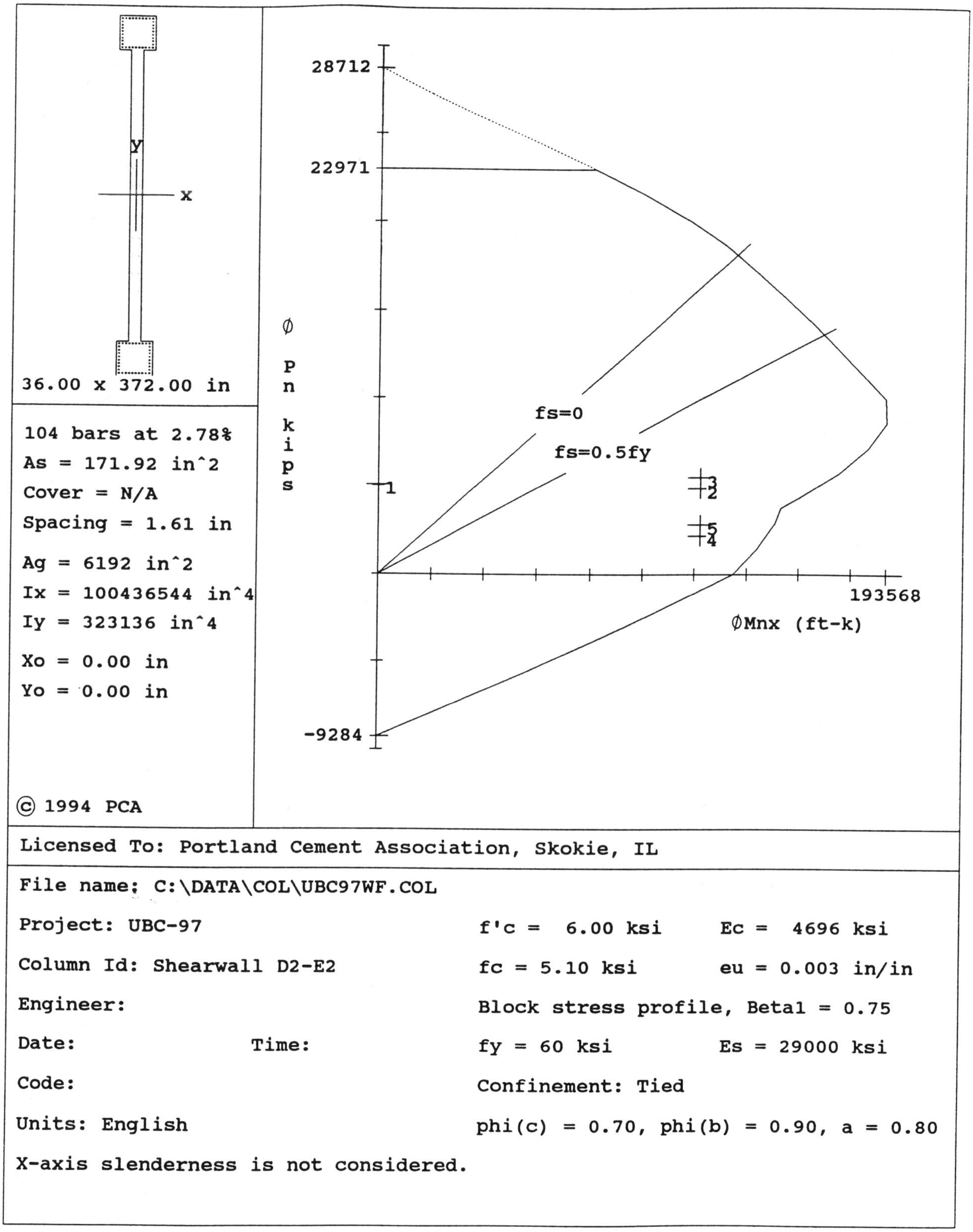

Figure 5-12 Design Strength Interaction Diagram for Shearwall D2-E2

The elastic displacement at the top of the wall (Δ_E) is equal to one-half of the design level response displacement (Δ_S), since 50% of the gross moment of inertia of the walls and beams were used in the analysis to obtain Δ_S. Thus, $\Delta_E = 0.5 \times 3.73 = 1.87$ in. (see Table 5-4). The corresponding bending moment (M_E) at the base of the wall is 111,052 ft-kips (see Table 5-3). From the analysis using the cracked section properties, the total displacement at the top of the wall (Δ_t) is equal to $\Delta_M = 14.4$ in. (see Table 5-4). Also,

$$\Delta_y = \frac{M'_n}{M_E}\Delta_E = \frac{206,307}{111,052} \times 1.87 = 3.5 \text{ in.}$$

The inelastic deflection at the top of the wall is:

$$\Delta_i = \Delta_t - \Delta_y = 14.4 - 3.5 = 10.9 \text{ in.}$$

Assuming that ℓ_p is equal to $0.5\ell_w = 0.5 \times 31 \times 12 = 186$ in., the total curvature demand is:

$$\phi_t = \frac{10.9}{[(133 \times 12) - (186/2)] \times 186} + \frac{0.003}{31 \times 12} = 3.90 \times 10^{-5} + 8.07 \times 10^{-6} = 4.71 \times 10^{-5}$$

Since ϕ_t is less than $0.003/c'_u = 0.003/46.7 = 6.42 \times 10^{-5}$, boundary zone details as defined in 1921.6.6.6 are not required. However, the boundary zone detail requirements of 1921.6.6.6 will be provided for illustrative purposes only.

Check minimum dimensions of the boundary zone:

$$\text{Minimum thickness} = \frac{\ell_u}{16} = \frac{(16 \times 12) - 24.5}{16} = 10.5 \text{ in.} < 36 \text{ in. O.K.}$$

$$\text{Minimum length} = 18 \text{ in.} < 36 \text{ in. O.K.}$$

The minimum area of confinement reinforcement (A_{sh}) is given in Eq. (21-10):

$$A_{sh} = \frac{0.09 s h_c f'_c}{f_{yh}}$$

Using the maximum allowable spacing of $6d_b = 6 \times 1.41 = 8.5$ in. or 6 in. (governs) and No. 5 hoops:

$$A_{sh} = \frac{0.09 \times 6 \times [36 - (2 \times 1.5) - 0.625] \times 6}{60} = 1.75 \text{ in.}^2$$

No. 5 hoops with crossties around every other longitudinal bar provide $A_{sh} = 6 \times 0.31 = 1.86$ in.^{2} Note that over the splice length of the vertical bars in the boundary zone, the hoops and crossties must be spaced at 4 in. on center. Also, the minimum area of vertical reinforcement in the boundary zone is $0.005 \times 36^2 = 6.48$ in.^{2} which is less than the area provided by 36-No. 14 bars. Figure 5-13 shows the reinforcement details for the shearwall.

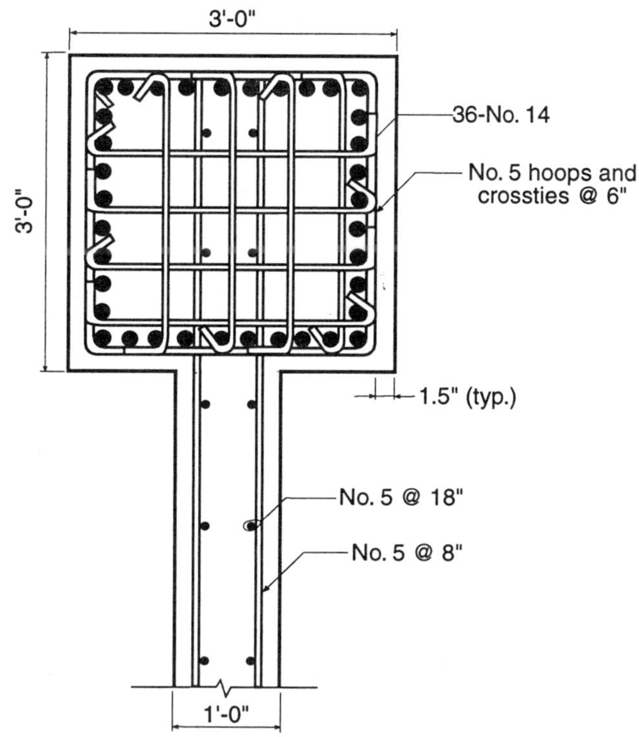

Figure 5-13 Reinforcement Details for Shearwall D2-E2

5.7 Reference

5.1 *PCACOL – Strength Design of Reinforced Concrete Column Sections*, Portland Cement Association, Skokie, IL, 1994.

6 DUAL SYSTEM

6.1 INTRODUCTION

A typical plan and elevation of a 12-story reinforced concrete office building is shown in Fig. 6-1. The computation of the seismic design forces according to the 1997 edition of the Uniform Building Code (UBC) is illustrated below. Typical beams, columns, and shearwalls are designed and detailed for the combined effects of gravity, seismic, and wind forces.

In a dual system, resistance to lateral loads is provided by shearwalls and moment-resisting frames. The two systems must be designed to resist the total base shear in proportion to their relative rigidities considering the interaction of the systems at all levels. Additionally, to safeguard against collapse, the moment-resisting frames must be designed to independently resist at least 25 percent of the design base shear.

It is assumed mainly for simplicity that the columns and shearwalls have constant cross-sections throughout the height of the building, and that the bases of the lowest story segments are fixed. The beams and slabs are also assumed to have the same dimensions at all floor levels. Although the member dimensions in this example are within the practical range, the structure itself is a hypothetical one, and has been chosen mainly for illustrative purposes.

6.2 DESIGN DATA

- Material Properties

  Concrete: f'_c = 4,000 psi, w_c = 145 pcf
  Reinforcement: f_y = 60,000 psi

- Service Loads

  Live loads: roof = 20 psf
  floor = 50 psf (Table 16-A)

  Superimposed dead loads: roof = 10 psf
  floors: partitions = 20 psf
  ceiling and mechanical = 10 psf

  Glass curtain wall = 8 psf
  Mechanical room on roof = 100 kips (located in the bay bounded by column lines B, C, 4, and 5)

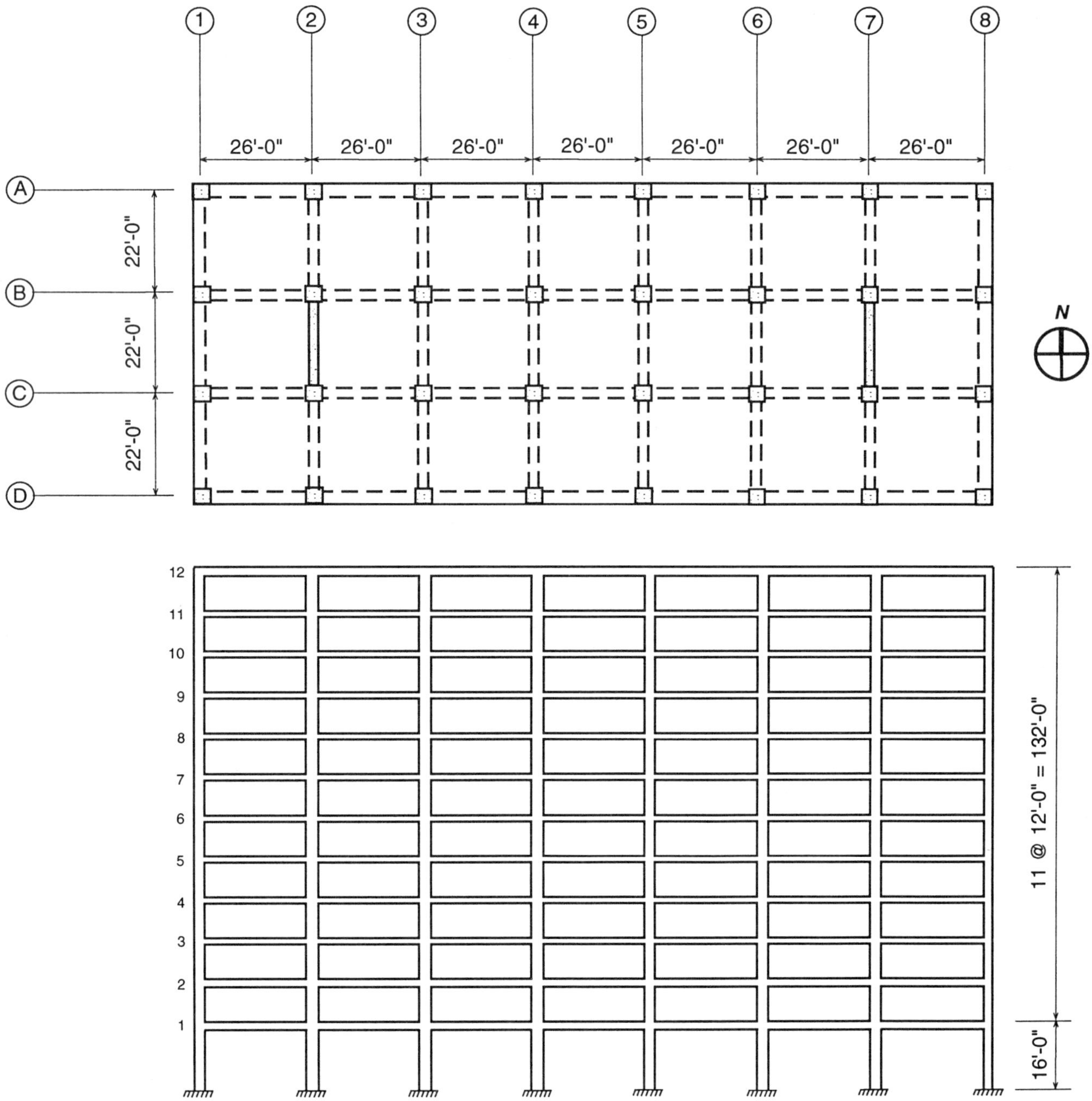

Edge columns: 22 x 22 in.
Interior columns: 26 x 26 in.
Beams: 20 x 24 in.
Slab: 8 in.
Walls: 18 in. web + 32 x 32 in. boundary elements

Figure 6-1 Typical Floor Plan and Elevation of Example Building

- Seismic Design Data

  Seismic Zone 4 ($Z = 0.40$; see Table 16-I)
  Seismic Source Type B (Table 16-U), with the site located more than 10 km from a known seismic source
  Soil Profile Type S_B (Table 16-J)
  For standard occupancy, Seismic Importance Factor (I) = 1.0 (Table 16-K)
  For the dual system (concrete shearwalls with a special moment-resisting frame) in the N-S direction, $R = 8.5$ (Table 16-N)
  For the special moment-resisting frame (SMRF) system in the E-W direction, $R = 8.5$ (Table 16-N)

- Wind Design Data

  Basic wind speed = 70 mph
  Exposure C
  For standard occupancy, Wind Importance Factor (I_w) = 1.0 (Table 16-K)

6.3 SEISMIC LOAD ANALYSIS

6.3.1 Seismic Forces

The seismic base shear (V) is determined below in both the N-S and E-W directions according to the static force procedure (1630.2). Note that this procedure is applicable for this regular building which is less than 240 ft in height (1629.8.3). The base shear in each direction is then distributed vertically to each floor level according to 1630.5.

Seismic Dead Load (*W*).

The dead loads for the roof level, a typical floor, and the first floor level are contained in Table 6-1. The total seismic dead load of the building (W) is equal to $1{,}961 + (10 \times 2{,}294) + 2{,}357 = 27{,}258$ kips.

Table 6-1 Dead Loads at Various Floor Levels (kips)

	Beams	Slab*	Vertical Elements**	Total
Roof	358	1,422†	181	1,961
Typical Floor	358	1,574	362	2,294
Floor Level 1	358	1,574	425	2,357

**includes the weight of the slab and the superimposed dead load*
***includes the weight of the curtain wall*
†includes the weight of the mechanical room

Approximate Period (*T*).

Method A in 1630.2.2 is used to compute the period (T) in each direction. In the N-S direction for the dual system:

Building height (h_n) = 148 ft

Numerical coefficient $C_t = 0.020$

Period $T = C_t\,(h_n)^{3/4} = 0.020 \times (148)^{3/4} = 0.849$ sec

Similarly, in the E-W direction for the SMRF:

Period $T = C_t\,(h_n)^{3/4} = 0.030 \times (148)^{3/4} = 1.273$ sec

Base Shear (*V*).

The total design base shear (V) in the N-S direction is determined from Eq. (30-4) using $T = 0.849$ sec:

$$V = \frac{C_v IW}{RT}$$

where $C_v = 0.40N_v = 0.40 \times 1.0 = 0.40$ for Seismic Zone 4, soil profile type S_B, and seismic source type B with the site located more than 10 km from the seismic source (see Tables 16-R and 16-T).

Therefore,

$$V = \frac{0.40 \times 1.0 \times 27{,}258}{8.5 \times 0.849} = 1{,}512 \text{ kips}$$

In the E-W direction:

$$V = \frac{0.40 \times 1.0 \times 27{,}258}{8.5 \times 1.273} = 1{,}010 \text{ kips}$$

The total design base shear need not exceed the value from Eq. (30-5):

$$V = \frac{2.5C_a IW}{R} = \frac{2.5 \times 0.40 \times 1.0 \times 27{,}258}{8.5} = 3{,}207 \text{ kips}$$

where $C_a = 0.40N_a = 0.40 \times 1.0 = 0.40$ for Seismic Zone 4, soil profile type S_B, and seismic source type B with the site located more than 10 km from the seismic source (see Tables 16-Q and 16-S).

Also, V shall not be less than the following values from Eqs. (30-6) and (30-7), the later equation being applicable for Seismic Zone 4:

$$V = \begin{cases} 0.11C_a IW = 0.11 \times 0.40 \times 1.0 \times 27{,}258 = 1{,}199 \text{ kips} \\ \dfrac{0.8ZN_v IW}{R} = \dfrac{0.8 \times 0.40 \times 1.0 \times 1.0 \times 27{,}258}{8.5} = 1{,}026 \text{ kips} \end{cases}$$

Thus, use $V = 1{,}512$ kips in the N-S direction and 1,199 kips in the E-W direction.

Lateral Forces (F_x).

The total base shear (V) is distributed over the height of the building in conformance with Eqs. (30-13), (30-14), and (30-15). The lateral force (F_x) at each floor level and the story shears (V_x) are contained in Tables 6-2 and 6-3 for the N-S and E-W directions, respectively.

Table 6-2 Seismic Forces and Story Shears in N-S Direction

Level	Story weight, w_x (kips)	Height, h_x (ft)	$w_x h_x$ (ft-kips)	Lateral Force, F_x* (kips)	Story Shear, V_x (kips)
12	1,961	148	290,228	277	277
11	2,294	136	311,984	201	478
10	2,294	124	284,456	183	661
9	2,294	112	256,928	165	826
8	2,294	100	229,400	148	974
7	2,294	88	201,872	130	1,104
6	2,294	76	174,344	112	1,216
5	2,294	64	146,816	95	1,311
4	2,294	52	119,288	77	1,388
3	2,294	40	91,760	59	1,447
2	2,294	28	64,232	41	1,448
1	2,357	16	37,712	24	1,512
Σ	27,258		2,209,020		

$$^*F_x = \frac{(V - F_t) w_x h_x}{\sum_{i=1}^{n} w_i h_i} \quad (\textit{see 1630.5})$$

Note that at the top of the building in the N-S direction, the required concentrated force F_t is equal to $0.07TV = 0.07 \times 0.849 \times 1{,}512 = 90$ kips which is less than $0.25V = 378$ kips (1630.5). This force must be added to the force F_n to obtain the total force at the roof level:

$$\text{Total force at roof level (N-S)} = F_n + F_t = \frac{(V - F_t) w_n h_n}{\sum_{i=1}^{n} w_i h_i} + F_t = \frac{(1{,}512 - 90) \times 290{,}228}{2{,}209{,}020} + 90 = 277 \text{ kips}$$

In the E-W direction, F_t is equal to $0.07 \times 1.273 \times 1{,}199 = 107$ kips which is less than $0.25V = 300$ kips, so that the total force at the roof level is:

$$\text{Total force at roof level (E-W)} = F_n + F_t = \frac{(1{,}199 - 107) \times 290{,}228}{2{,}209{,}020} + 107 = 250 \text{ kips}$$

Table 6-3 Seismic Forces and Story Shears in E-W Direction

Level	Story weight, w_x (kips)	Height, h_x (ft)	$w_x h_x$ (ft-kips)	Lateral Force, F_x* (kips)	Story Shear, V_x (kips)
12	1,961	148	290,228	250	250
11	2,294	136	311,984	154	404
10	2,294	124	284,456	141	545
9	2,294	112	256,928	127	672
8	2,294	100	229,400	113	785
7	2,294	88	201,872	100	885
6	2,294	76	174,344	86	971
5	2,294	64	146,816	73	1,044
4	2,294	52	119,288	59	1,103
3	2,294	40	91,760	45	1,148
2	2,294	28	64,232	32	1,180
1	2,357	16	37,712	19	1,199
Σ	27,258		2,209,020		

$$^*F_x = \frac{(V - F_t)w_x h_x}{\sum_{i=1}^{n} w_i h_i} \quad \text{(see 1630.5)}$$

6.3.2 Analysis Results

A three-dimensional analysis of the building was performed in both the N-S and E-W directions for the seismic forces in Tables 6-2 and 6-3, respectively, using *SAP90* (Ref. 6.1). In the model, rigid diaphragms were assigned at each floor level, and rigid-end offsets were defined at the ends of each horizontal member so that results were automatically obtained at the faces of each support. For the beams, the portion of the adjoining slab defined in 1913.2.4 was included in the analysis as being part of the beam. In order to satisfy the modeling requirements of 1630.1.2, the stiffness properties of the members were input assuming cracked sections. In lieu of a more accurate analysis, the provisions of 1910.11.1 were used, i.e., the moment of inertia for the beams was taken as $0.35I_g$ where I_g is the gross moment of inertia of the section; for the columns, it was taken as $0.70I_g$; and for the cracked wall sections, it was taken as $0.35I_g$. Additionally, the modulus of elasticity was computed from 1908.5.1, and the gross area of the members was used in the analysis per 1910.11.1. In accordance with 1630.6, the mass at each level was assumed to be displaced from the calculated center of mass a distance equal to 5 percent of the building dimension at that level perpendicular to the direction of the force under consideration.

In a dual system, an additional safeguard against total collapse is provided by requiring that the moment-resisting frames be capable of resisting 25% of the base shear without the benefit of the shearwalls (1629.6.5). Thus, the structure was also analyzed in the N-S direction using 25% of the base shear without the shearwalls present, including torsional effects.

A summary of the results of the analysis for the interior frame along column line 6 for seismic forces in the N-S direction can be found in Fig. 6-2. Note that italic text denotes results with 25% of the design base shear applied to the frames. The reactions for the shearwall on column line 7 due to seismic forces in the N-S direction are contained in Table 6-4.

Table 6-4 Bending Moments and Shear Forces for the Shearwall on Column Line 7 Due to Seismic Forces in the N-S Direction

Story	Bending Moment (ft-kips)		Shear Force (kips)
	Top	Bottom	
12	873	-1,898	-84
11	2,939	-2,099	70
10	3,149	-1,236	160
9	2,324	615	244
8	506	3,354	322
7	-2,228	6,981	396
6	-5,838	11,369	462
5	-10,309	16,603	524
4	-15,598	22,609	584
3	-21,743	29,458	644
2	-28,733	37,120	696
1	-36,617	49,005	774

6.3.3 Story Drift Limitation and P-Δ Effects

Table 6-5 contains the displacements (Δ_S) obtained from the 3-D static, elastic analyses using the design seismic forces in both the N-S and E-W directions, including accidental torsional effects. Note that for the E-W direction, the seismic design forces used to determine the displacements were from Eq. (30-4), disregarding the limitations of Eq. (30-6) (1630.10.3). The table also contains the maximum inelastic response displacements (Δ_M) which are computed from Eq. (30-17):

$$\Delta_M = 0.7 R \Delta_S$$

where R is equal to 8.5 in both directions. The interstory drifts computed from the Δ_S and Δ_M are also contained in the table.

According to 1630.10.2, the calculated story drifts using Δ_M shall not exceed 0.020 times the story height for structures having a fundamental period of 0.7 seconds or greater. For the 12 ft story height, the maximum drift is equal to $0.020 \times 12 \times 12 = 2.88$ in., and for the 16 ft story height it is equal to 3.84 in. It is clear from Table 6-5 that for all stories in both directions, the lateral drifts obtained from the prescribed lateral forces are less than the limiting values.

According to 1630.1.3, P-Δ effects need not be considered for Seismic Zones 3 and 4 when the calculated story drift using Δ_S is less than or equal to $0.02h_s/R$, which is $0.02 \times 16 \times 12/8.5 = 0.45$ in. for the first floor and $0.02 \times 12 \times 12/8.5 = 0.34$ in. for a typical floor. It can be seen from Table 6-5 that P-Δ effects must be considered for the elements below level 7 for analysis in the E-W direction only.

Level			
	77	78	82
12	*13*	*12*	*23*
	115	116	120
11	*27*	*27*	*38*
	117	119	123
10	*40*	*40*	*52*
	122	124	128
9	*54*	*54*	*65*
	126	127	132
8	*65*	*66*	*77*
	128	129	133
7	*76*	*76*	*87*
	127	129	133
6	*85*	*85*	*96*
	124	125	129
5	*93*	*93*	*103*
	116	117	121
4	*99*	*100*	*109*
	104	105	108
3	*104*	*105*	*113*
	87	88	90
2	*108*	*109*	*115*
	64	65	66
1	*104*	*105*	*109*

Bending Moments - Beams (ft-kips)

79	*13*	155	*34*
59	*6*	118	*19*
57	*21*	117	*45*
57	*13*	115	*32*
62	*27*	124	*59*
60	*21*	120	*46*
64	*33*	129	*71*
62	*27*	126	*60*
65	*38*	131	*81*
65	*33*	130	*71*
65	*43*	130	*90*
66	*38*	132	*82*
63	*47*	127	*98*
66	*43*	132	*91*
59	*50*	119	*104*
64	*47*	129	*99*
53	*52*	107	*109*
61	*51*	122	*105*
44	*54*	89	*111*
54	*53*	109	*109*
34	*55*	68	*112*
47	*58*	95	*119*
17	*46*	35	*92*
44	*88*	86	*172*

Bending Moments - Columns (ft-kips)

8	*1*	5	*1*
19	*4*	9	*1*
30	*7*	14	*2*
42	*12*	19	*2*
54	*18*	24	*3*
66	*25*	29	*4*
78	*33*	34	*4*
90	*42*	38	*5*
101	*51*	43	*6*
111	*60*	46	*6*
119	*70*	49	*6*
125	*80*	51	*7*

Axial Forces - Columns (kips)

Note: *Italic* text denotes results with 25% of design base shear applied to frames.

Figure 6-2 Analysis Results Along Column Line 6 for Seismic Forces in the N-S Direction

Table 6-5 Lateral Displacements and Interstory Drifts due to Seismic Forces (in.)

Story	N-S				E-W			
	Δ_S	Drift	Δ_M	Drift	Δ_S	Drift	Δ_M	Drift
12	2.39	0.20	14.22	1.19	4.10	0.11	24.40	0.66
11	2.19	0.22	13.03	1.31	3.99	0.17	23.74	1.01
10	1.97	0.22	11.72	1.31	3.82	0.22	22.73	1.31
9	1.75	0.23	10.41	1.37	3.60	0.28	21.42	1.67
8	1.52	0.24	9.04	1.43	3.32	0.32	19.75	1.90
7	1.28	0.23	7.62	1.37	3.00	0.35	17.85	2.08
6	1.05	0.23	6.25	1.37	2.65	0.40	15.77	2.38
5	0.82	0.22	4.88	1.31	2.25	0.42	13.39	2.50
4	0.60	0.19	3.57	1.13	1.83	0.44	10.89	2.62
3	0.41	0.17	2.44	1.01	1.39	0.46	8.27	2.74
2	0.24	0.14	1.43	0.83	0.93	0.46	5.53	2.74
1	0.10	0.10	0.60	0.60	0.47	0.47	2.80	2.80

6.4 WIND LOAD ANALYSIS

Tables 6-6 and 6-7 contain the wind loads in the N-S and E-W directions, respectively, computed according to 1609 with the basic wind speed equal to 70 mph, exposure C, and a wind importance factor (I_w) equal to 1.0.

Table 6-6 Design Wind Forces in N-S Direction (for entire structure)

Level	Height h_x (ft)	Tributary Height (ft)	Windward			Leeward			Total Force (kips)
			C_e	Pressure* (psf)	Force** (kips)	C_e	Pressure* (psf)	Force** (kips)	
12	148	6	1.75	17.6	19.4	1.75	11.0	12.2	31.6
11	136	12	1.72	17.3	38.2	1.75	11.0	24.3	62.6
10	124	12	1.68	16.9	37.4	1.75	11.0	24.3	61.7
9	112	12	1.65	16.6	36.7	1.75	11.0	24.3	61.0
8	100	12	1.61	16.2	35.8	1.75	11.0	24.3	60.1
7	88	12	1.56	15.7	34.7	1.75	11.0	24.3	59.0
6	76	12	1.51	15.2	33.6	1.75	11.0	24.3	57.9
5	64	12	1.45	14.6	32.2	1.75	11.0	24.3	56.5
4	52	12	1.38	13.9	30.7	1.75	11.0	24.3	55.0
3	40	12	1.31	13.2	29.1	1.75	11.0	24.3	53.4
2	28	12	1.21	12.2	26.9	1.75	11.0	24.3	51.2
1	16	14	1.07	10.8	27.8	1.75	11.0	28.4	56.2

*Windward pressure = $C_eC_qq_sI_w = C_e \times 0.8 \times 12.6 \times 1.0 = 10.08C_e$ psf

Leeward pressure = $C_eC_qq_sI_w = 1.75 \times 0.5 \times 12.6 \times 1.0 = 11$ psf

**Force = pressure × tributary height × 183.83/1,000

A summary of the results of the analysis for the interior frame along column line 6 for wind forces in the N-S direction is contained in Fig. 6-3. Table 6-8 contains the reactions for the shearwall on column line 7 due to the wind loads in the N-S direction. The lateral displacements in both directions are provided in Table 6-9.

Table 6-7 Design Wind Forces in E-W Direction (for entire structure)

Level	Height h_x (ft)	Tributary Height (ft)	Windward			Leeward			Total Force (kips)
			C_e	Pressure* (psf)	Force** (kips)	C_e	Pressure* (psf)	Force** (kips)	
12	148	6	1.75	17.6	7.2	1.75	11.0	4.5	11,7
11	136	12	1.72	17.3	14.1	1.75	11.0	9.0	23.1
10	124	12	1.68	16.9	13.8	1.75	11.0	9.0	22.8
9	112	12	1.65	16.6	13.5	1.75	11.0	9.0	22.5
8	100	12	1.61	16.2	13.2	1.75	11.0	9.0	22.2
7	88	12	1.56	15.7	12.8	1.75	11.0	9.0	21.8
6	76	12	1.51	15.2	12.4	1.75	11.0	9.0	21.4
5	64	12	1.45	14.6	11.9	1.75	11.0	9.0	20.9
4	52	12	1.38	13.9	11.3	1.75	11.0	9.0	20.3
3	40	12	1.31	13.2	10.7	1.75	11.0	9.0	19.7
2	28	12	1.21	12.2	9.9	1.75	11.0	9.0	18.9
1	16	14	1.07	10.8	10.2	1.75	11.0	10.5	20.7

**Windward pressure = $C_eC_qq_sI_w = C_e \times 0.8 \times 12.6 \times 1.0 = 10.08C_e$ psf*

Leeward pressure = $C_eC_qq_sI_w = 1.75 \times 0.5 \times 12.6 \times 1.0 = 11$ psf

***Force = pressure × tributary height × 67.83/1,000*

Table 6-8 Bending Moments and Shear Forces for the Shearwall on Column Line 7 Due to Wind Forces in the N-S Direction

Story	Bending Moment (ft-kips)		Shear Force (kips)
	Top	Bottom	
12	207	-725	-44
11	991	-1,043	-4
10	1,312	-1,015	24
9	1,291	-679	50
8	983	-32	78
7	363	905	106
6	-594	2,212	134
5	-1,902	3,842	164
4	-3,560	5,878	192
3	-5,607	8,317	226
2	-8,089	11,209	260
1	-11,069	15,933	306

Table 6-9 Lateral Displacements due to Wind Forces (in.)

Story	N-S	E-W
12	0.75	0.77
11	0.69	0.76
10	0.63	0.74
9	0.57	0.71
8	0.50	0.67
7	0.43	0.62
6	0.36	0.56
5	0.29	0.49
4	0.22	0.41
3	0.15	0.32
2	0.09	0.22
1	0.04	0.11

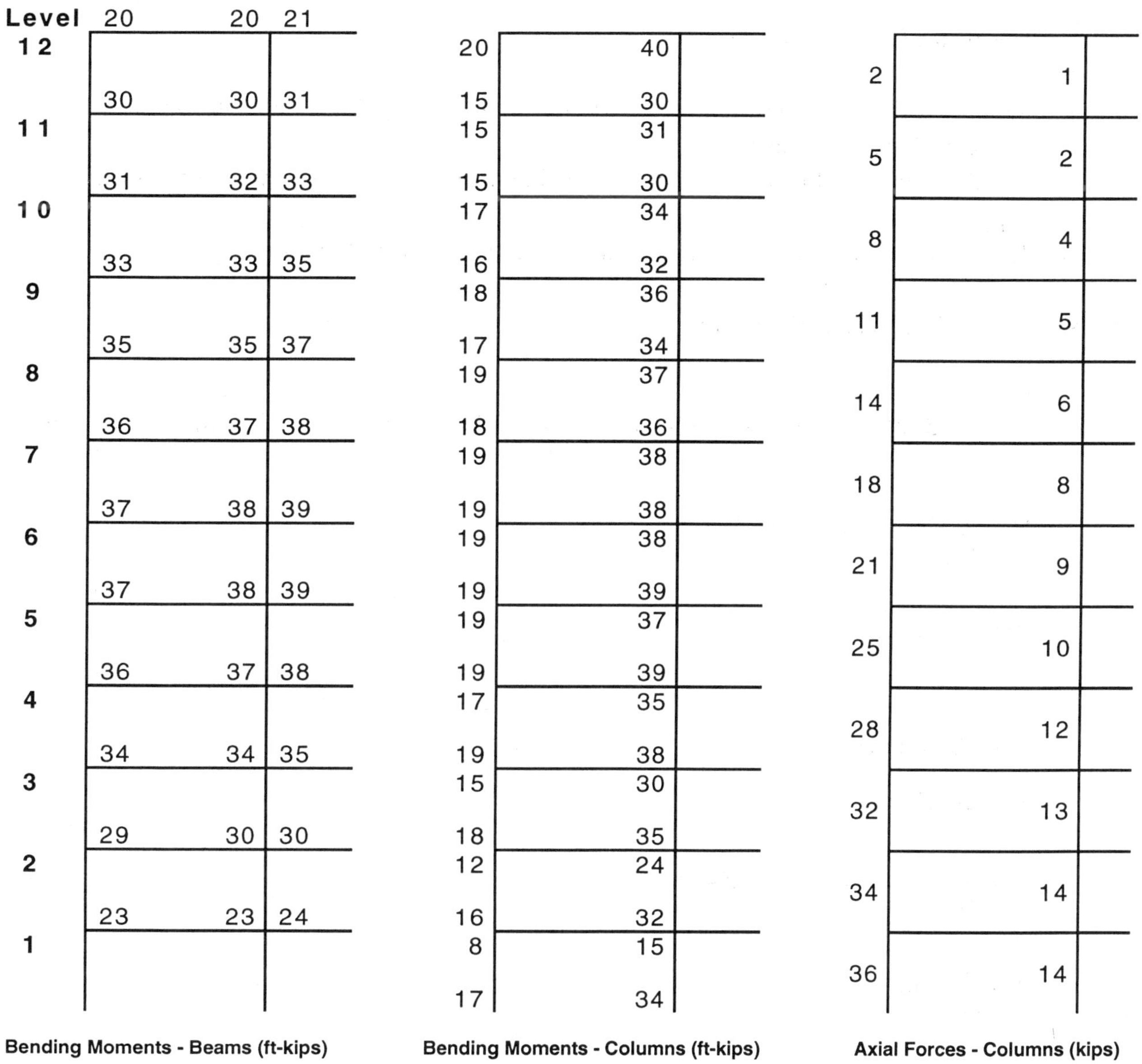

Figure 6-3 Analysis Results Along Column Line 6 for Wind Forces in the N-S Direction

6.5 GRAVITY LOAD ANALYSIS

6.5.1 Slab and Beams

The Direct Design Method (1913.6) is used to compute the gravity load moments in the slab and beams on a typical floor along column line 6. Note that this method is approximate and can only be used if certain geometric and loading criteria are satisfied (see 1913.6.1). In this case, all of the criteria are satisfied, including 1913.6.1.6:

$$5.0 \geq \frac{\alpha_1 \ell_2^2}{\alpha_2 \ell_1^2} \geq 0.2$$

where $\alpha_1, \alpha_2 = \alpha$ in direction of ℓ_1, ℓ_2

$\alpha = I_b / I_s$

I_b = moment of inertia of beam, including portion of adjoining slab defined in 1913.2.4

I_s = moment of inertia of slab

ℓ_1 = length of span in direction moments are being determined, measured center-to-center of supports

ℓ_2 = length of span transverse to ℓ_1, measured center-to-center of supports

For the N-S interior beams (see Fig. 6-4a):

$$I_b = \frac{1}{12}(52)(8)^3 + (52 \times 8)(20 - 14.78)^2 + \frac{1}{12}(20)(16)^3 + (20 \times 16)(14.78 - 8)^2 = 35{,}091 \text{ in.}^4$$

$$I_s = \frac{1}{12}(26 \times 12)(8)^3 = 13{,}312 \text{ in.}^4$$

$$\alpha = \frac{I_b}{I_s} = 2.6$$

Similarly, for the E-W interior beams:

$$I_b = 35{,}091 \text{ in.}^4$$

$$I_s = \frac{1}{12}(22 \times 12)(8)^3 = 11{,}264 \text{ in.}^4$$

$$\alpha = \frac{I_b}{I_s} = 3.1$$

For the E-W edge beams (see Fig. 6-4b):

$$I_b = \frac{1}{12}(36)(8)^3 + (36 \times 8)(20 - 13.68)^2 + \frac{1}{12}(20)(16)^3 + (20 \times 16)(13.68 - 8)^2 = 30{,}190 \text{ in.}^4$$

$$I_s = \frac{1}{12}(11.92 \times 12)(8)^3 = 6{,}101 \text{ in.}^4$$

$$\alpha = \frac{I_b}{I_s} = 5.0$$

Thus, for an interior panel with the direction of analysis in the N-S direction:

$$\frac{\alpha_1 \ell_2^2}{\alpha_2 \ell_1^2} = \frac{2.6 \times 26^2}{3.1 \times 22^2} = 1.2, \; 5.0 > 1.2 > 0.2 \quad \text{O.K.}$$

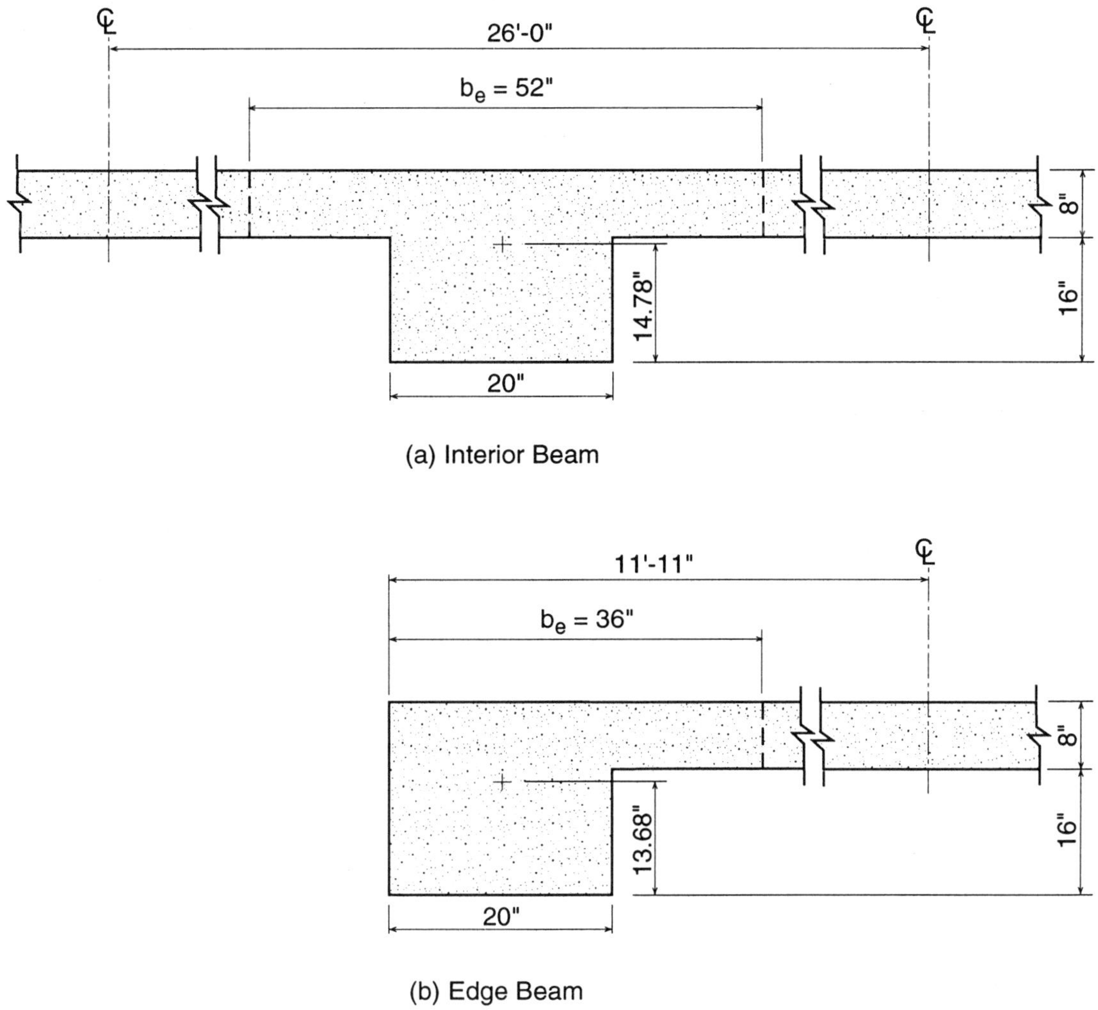

Figure 6-4 Section Properties of Interior and Edge Beams for Stiffness Computations

For an edge panel:

$$\frac{\alpha_1 \ell_2^2}{\alpha_2 \ell_1^2} = \frac{2.6 \times 26^2}{5.0 \times 22^2} = 0.7, \; 5.0 > 0.7 > 0.2 \quad \text{O.K.}$$

The bending moments are determined using the dead and live loads per square foot of panel:

$$\text{Dead load} = \frac{8}{12}(145) + 30 + \frac{16 \times 20}{144}(145) \times \frac{1}{26} = 139 \text{ psf}$$

$$\text{Live load} = 50 \text{ psf}$$

Using Eq. (13-3), the total static service dead and live load moments for an end span in the N-S direction are computed as follows:

$$M_o = \frac{w\ell_2\ell_n^2}{8}$$

$$(M_o)_{DL} = \frac{0.139 \times 26 \times 20^2}{8} = 180.7 \text{ ft - kips}$$

$$(M_o)_{LL} = \frac{0.05 \times 26 \times 20^2}{8} = 65.0 \text{ ft - kips}$$

Similarly, for an interior span, ℓ_n = 19.83 ft so that $(M_o)_{DL}$ = 177.7 ft-kips and $(M_o)_{LL}$ = 63.9 ft-kips.

These moments must be divided into positive and negative moments, and then into column strip and middle strip moments according to the distribution factors given in 1913.6.3, 1913.6.4, and 1913.6.6. The column strip moments are further divided into beam and slab moments. The beams are designed for 85% of the column strip moments since $\alpha_1\ell_2/\ell_1$ = (2.6 × 26/22) = 3.1 > 1.0 (1913.6.5.1). The distribution factors for a two-way beam-supported slab are given in Table 6-10.

It is important to note that the distribution factors contained in Table 6-10 for the exterior negative moments in the column strip are based on β_t greater than or equal to 2.5, where β_t is the ratio of the torsional stiffness of the edge beam to the flexural stiffness of the slab (see 1913.0). In this case, the torsional constant of the edge beam (C) is the larger value from the two cases shown in Fig. 6-5 (see 1913.0 for the definition of C):

- Case A:

$$C_1 = \left(1 - 0.63\frac{20}{24}\right)\left(\frac{20^3 \times 24}{3}\right) = 30{,}400 \text{ in.}^4$$

$$C_2 = \left(1 - 0.63\frac{8}{16}\right)\left(\frac{8^3 \times 16}{3}\right) = 1{,}871 \text{ in.}^4$$

$$C = C_1 + C_2 = 32{,}271 \text{ in.}^4 \quad \text{(governs)}$$

- Case B:

$$C_1 = \left(1 - 0.63\frac{8}{36}\right)\left(\frac{8^3 \times 36}{3}\right) = 5{,}284 \text{ in.}^4$$

$$C_2 = \left(1 - 0.63\frac{16}{20}\right)\left(\frac{16^3 \times 20}{3}\right) = 13{,}544 \text{ in.}^4$$

$$C = C_1 + C_2 = 18{,}828 \text{ in.}^4$$

Table 6-10 Design Moments at Face of Support for Two-Way Beam-Supported Slab

Span Ratio	Moments		End Span			Interior Span	
			Exterior Negative	Positive	Interior Negative	Positive	Interior Negative
ℓ_2/ℓ_1	Total Moment		$0.16M_o$	$0.57M_o$	$0.70M_o$	$0.35M_o$	$0.65M_o$
0.5	Column Strip	Beam	$0.12M_o$	$0.43M_o$	$0.54M_o$	$0.27M_o$	$0.50M_o$
		Slab	$0.02M_o$	$0.08M_o$	$0.09M_o$	$0.05M_o$	$0.09M_o$
	Middle Strip		$0.02M_o$	$0.06M_o$	$0.07M_o$	$0.03M_o$	$0.06M_o$
1.0	Column Strip	Beam	$0.10M_o$	$0.37M_o$	$0.45M_o$	$0.22M_o$	$0.42M_o$
		Slab	$0.02M_o$	$0.06M_o$	$0.08M_o$	$0.04M_o$	$0.07M_o$
	Middle Strip		$0.04M_o$	$0.14M_o$	$0.17M_o$	$0.09M_o$	$0.16M_o$
2.0	Column Strip	Beam	$0.06M_o$	$0.22M_o$	$0.27M_o$	$0.14M_o$	$0.25M_o$
		Slab	$0.01M_o$	$0.04M_o$	$0.05M_o$	$0.02M_o$	$0.04M_o$
	Middle Strip		$0.09M_o$	$0.31M_o$	$0.38M_o$	$0.19M_o$	$0.36M_o$

Notes: (1) moments based on $\alpha_1\ell_2/\ell_1 \geq 1.0$ and $\beta_t \geq 2.5$
(2) interpolate between values shown for different ℓ_2/ℓ_1 ratios.
(3) all negative moments are at face of support.
(4) concentrated loads applied directly to beams must be accounted for separately.

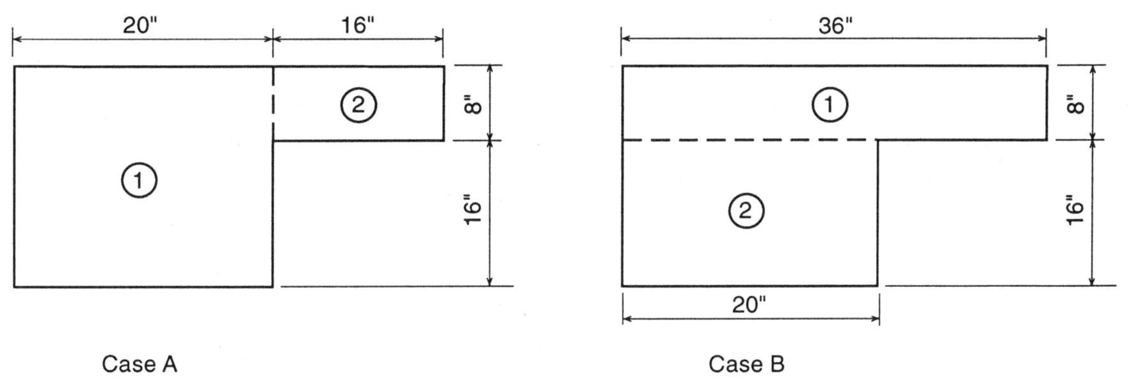

Figure 6-5 Component Rectangles to be Used in Computation of Torsional Constant (C)

In determining the moment of inertia of the slab (I_s), the width of the slab must equal the center-to-center length of the edge beam:

$$I_s = \frac{1}{12} \times (26 \times 12) \times (8)^3 = 13{,}312 \text{ in.}^4$$

Thus,

$$\beta_t = \frac{C}{2I_s} = \frac{32{,}271}{2 \times 13{,}312} = 1.2$$

Since β_t is less than 2.5, the distribution factors in Table 6-10 for the exterior negative moments in the column strip cannot be used. For the case when $\alpha_1\ell_2/\ell_1 > 1.0$, the following equations may be used to determine the percentage of total negative and positive bending moments to be resisted by the column strip (see Ref. 6.2):

Percentage of negative moment at exterior support to be resisted by the column strip

$$= 100 - 10\beta_t + 12\beta_t\left(1 - \frac{\ell_2}{\ell_1}\right) = 100 - (10 \times 1.2) + (12 \times 1.2)\left(1 - \frac{26}{22}\right) = 85.4\%$$

Percentage of negative moment at interior support to be resisted by the column strip

$$= 75 + 30\left(1 - \frac{\ell_2}{\ell_1}\right) = 75 + 30\left(1 - \frac{26}{22}\right) = 69.6\%$$

Percentage of positive moment to be resisted by the column strip

$$= 60 + 30\left(1.5 - \frac{\ell_2}{\ell_1}\right) = 60 + 30\left(1.5 - \frac{26}{22}\right) = 69.6\%$$

Thus, the total column strip moment at the exterior support in the end span is equal to $0.854 \times 0.16M_o = 0.14M_o$, with $0.85 \times 0.14M_o = 0.12M_o$ transferred to the beam and $0.02M_o$ transferred to the slab. The middle strip moment is $0.02M_o$ at this location. The moments at other locations along the span can be determined in a similar fashion. Table 6-11 contains the service gravity load moments for the slab and beams of an end span and an interior span on a typical floor.

Table 6-11 Service Dead and Live Load Moments (ft-kips) for Slab and Beams on a Typical Floor Along Column Line 6

			Moment	M_{DL}	M_{LL}
END SPAN					
Column Strip	Ext. Negative	Beam	$0.12M_o$	-21.7	-7.8
		Slab	$0.02M_o$	-3.6	-1.3
	Positive	Beam	$0.34M_o$	61.4	22.1
		Slab	$0.06M_o$	10.8	3.9
	Int. Negative	Beam	$0.41M_o$	-74.1	-26.7
		Slab	$0.08M_o$	-14.5	-5.2
Middle Strip	Ext. Negative		$0.02M_o$	-3.6	-1.3
	Positive		$0.17M_o$	30.7	11.1
	Int. Negative		$0.21M_o$	-38.0	-13.7
INTERIOR SPAN					
Column Strip	Positive	Beam	$0.21M_o$	37.3	13.4
		Slab	$0.03M_o$	5.3	1.9
	Int. Negative	Beam	$0.39M_o$	-69.3	-24.9
		Slab	$0.06M_o$	-10.7	-3.8
Middle Strip	Positive		$0.11M_o$	19.6	7.0
	Int. Negative		$0.20M_o$	-35.5	-12.8

6.5.2 Columns and Shearwalls

The service axial loads for an interior and edge column along column line 6 are given in Tables 6-12 and 6-13, respectively. Table 6-14 contains the service axial loads for a shearwall. In all cases, the live load reduction factor was computed in accordance with 1607.5.

Table 6-12 Service Dead and Live Load Axial Forces for an Interior Column Along Column Line 6

Story	Dead Load* (psf)	Live Load (psf)	Supported Area (ft^2)	Reduced Live Load** (psf)	Live Load (kips)	Cumulative Dead Load† (kips)	Cumulative Live Load (kips)
12	106.67	20	572	13.25	8	83	8
11	126.67	50	1,144	20.00	11	177	19
10	126.67	50	1,716	20.00	11	271	30
9	126.67	50	2,288	20.00	11	365	42
8	126.67	50	2,860	20.00	11	459	53
7	126.67	50	3,432	20.00	11	553	65
6	126.67	50	4,004	20.00	11	648	76
5	126.67	50	4,576	20.00	11	742	88
4	126.67	50	5,148	20.00	11	836	99
3	126.67	50	5,720	20.00	11	930	111
2	126.67	50	6,292	20.00	11	1,024	122
1	126.67	50	6,864	20.00	11	1,121	133

**includes the weight of the slab and the superimposed dead load*

***see 1607.5*

†includes beam stem weight of 0.32 kips/ft and column weight of 0.68 kips/ft

Table 6-13 Service Dead and Live Load Axial Forces for an Edge Column Along Column Line 6

Story	Dead Load* (psf)	Live Load (psf)	Supported Area (ft^2)	Reduced Live Load** (psf)	Live Load (kips)	Cumulative Dead Load† (kips)	Cumulative Live Load (kips)
12	106.67	20	310	17.44	5	50	5
11	126.67	50	620	31.21	10	105	15
10	126.67	50	929	20.00	6	161	21
9	126.67	50	1,239	20.00	6	217	27
8	126.67	50	1,549	20.00	6	273	34
7	126.67	50	1,859	20.00	6	328	40
6	126.67	50	2,169	20.00	6	384	46
5	126.67	50	2,479	20.00	6	440	52
4	126.67	50	2,788	20.00	6	496	58
3	126.67	50	3,098	20.00	6	551	65
2	126.67	50	3,408	20.00	6	607	71
1	126.67	50	3,718	20.00	6	665	77

**includes the weight of the slab and the superimposed dead load*

***see 1607.5*

†includes beam stem weight of 0.32 kips/ft, column weight of 0.49 kips/ft, and window weight of 0.21 kips/ft

Table 6-14 Service Dead and Live Load Axial Forces for a Shearwall

Story	Dead Load* (psf)	Live Load (psf)	Supported Area (ft^2)	Reduced Live Load** (psf)	Live Load (kips)	Cumulative Dead Load† (kips)	Cumulative Live Load (kips)
12	106.67	20	1,144	12.00	14	215	14
11	126.67	50	2,288	20.00	23	453	37
10	126.67	50	3,432	20.00	23	691	59
9	126.67	50	4,576	20.00	23	928	82
8	126.67	50	5,720	20.00	23	1,166	105
7	126.67	50	6,864	20.00	23	1,404	128
6	126.67	50	8,008	20.00	23	1,642	151
5	126.67	50	9,152	20.00	23	1,880	174
4	126.67	50	10,296	20.00	23	2,117	197
3	126.67	50	11,440	20.00	23	2,355	220
2	126.67	50	12,584	20.00	23	2,593	243
1	126.67	50	13,728	20.00	23	2,856	265

**includes the weight of the slab and the superimposed dead load*

***see 1607.5*

†includes beam stem weight of 0.32 kips/ft and wall weight of 6.3 kips/ft

6.6 DESIGN FOR COMBINED LOAD EFFECTS

6.6.1 Load Combinations

The basic load combinations for the strength design method are given in 1612.2.1. According to the exceptions at the end of that section, the load combinations of 1909.2 are applicable for concrete where the load combinations do not include seismic forces. Also, the load combinations of 1612.2.1 which include seismic forces must be multiplied by 1.1 for concrete (see Sect. 2.2 for discussion on the 1.1 factor). Thus, the following load combinations are applicable to concrete members subject to the combined effects of dead (D), live (L), wind (W), and earthquake (E) forces:

(1) $1.4D + 1.7L$

(2) $0.75(1.4D + 1.7L + 1.7W)$

(3) $0.9D + 1.3W$

(4) $1.1(1.2D + f_1L + 1.0E) = 1.32D + 1.1f_1L + 1.1E$

(5) $1.1(0.9D + 1.0E) = 0.99D + 1.1E$

where f_1 = 1.0 for floors in places of public assembly, for live loads in excess of 100 psf, and for garage live loads
= 0.5 for other live loads

The earthquake force (E) is determined from Eq. (30-1):

$$E = \rho E_h + E_v$$

where E_h = earthquake force due to the base shear (V)

E_v = the load effect resulting from the vertical component of the earthquake motion
$= 0.5C_aID = 0.5 \times 0.40 \times 1.0 \times D = 0.2D$ in this case

ρ = reliability/redundancy factor (see Eq. 30-3):

ρ = reliability/redundancy factor (see Eq. 30-3):

$$1.0 \le \rho = 2 - \frac{20}{r_{max}\sqrt{A_B}} \le 1.5$$

r_{max} = maximum element-story shear ratio

A_B = ground floor area of the structure, ft^2

According to 1630.1.1, the maximum element-story shear ratio (r_{max}) is defined as the largest of the element-story shear ratios (r_i) which occurs in any of the story levels at or below the two-thirds height level of the building. For a given direction of loading, r_i is the ratio of the design story shear in the most heavily loaded single element divided by the total design story shear.

In this case, the most heavily loaded single element for seismic forces in the N-S direction is a shearwall. Table 6-15 contains the shear forces in the shearwall which include the effects from accidental torsion (see Table 6-4), the story shear (see Table 6-2), and the values of r_i for the story levels below the two-thirds height level of the building. The maximum r_i occurs at the lowest level, so that r_{max} is equal to 0.51. Therefore,

$$\rho = 2 - \frac{20}{0.51\sqrt{183.83 \times 67.83}} = 1.65$$

Table 6-15 Element-Story Shear Ratios (r_i) for Seismic Forces in N-S Direction

Story	Shear Force in Shearwall (kips)	Story Shear (kips)	r_i = Shear Force/Story Shear
7	396	1,104	0.36
6	462	1,216	0.38
5	524	1,311	0.40
4	584	1,388	0.42
3	644	1,447	0.45
2	696	1,488	0.47
1	774	1,512	0.51

For dual systems, ρ need not exceed 80 percent of the value determined above (1630.1.1); thus, $\rho = 0.8 \times 1.65 = 1.3$. Therefore, the seismic force (E) is:

$$E = 1.3E_h + 0.2D$$

Substituting E and $f_1 = 0.5$ (live load < 100 psf) into load combinations 4 and 5 above results in the following (see Sect. 2.4.1 regarding the use of $0.2D$ in Eq. (5)):

$$(4)\ 1.32D + (1.1 \times 0.5)L + 1.1E = 1.32D + 0.55L + 1.1(1.3E_h + 0.2D) = 1.54D + 0.55L + 1.43E_h$$

$$(5)\ 0.99D + 1.1E = 0.99D + 1.1(1.3E_h - 0.2D) = 0.77D + 1.43E_h$$

Note that the values of E_h are given in Fig. 6-2 for the beams and columns along column line 6, and in Table 6-4 for the shearwall along column line 7. Also note that the special seismic load combinations of 1612.4 are not applicable in this example.

6.6.2 Beam and Slab Design

A summary of the slab and beam bending moments at the sixth floor level, based on the applicable load combinations, is contained in Table 6-16.

Table 6-16 Summary of Slab and Beam Design Moments Along Column Line 6 at Floor Level 6

No.	Load Case	Location	End Span Moments (ft-kips)			Interior Span Moments (ft-kips)		
			Column Strip		Middle Strip	Column Strip		Middle Strip
			Slab	Beam	Slab	Slab	Beam	Slab
	Dead (D)	Ext. Neg.	-3.6	-21.7	-3.6	—	—	—
		Positive	10.8	61.4	30.7	5.3	37.3	19.6
		Int. Neg.	-14.5	-74.1	-38.0	-10.7	-69.3	-35.5
	Live (L)	Ext. Neg.	-1.3	-7.8	-1.3	—	—	—
		Positive	3.9	22.1	11.1	1.9	13.4	7.0
		Int. Neg.	-5.2	-26.7	-13.7	-3.8	-24.9	-12.8
	Earthquake (E_h)	Ext. Neg.	—	±127.0	—	—	—	—
		Positive	—	—	—	—	—	—
		Int. Neg.	—	±129.0	—	—	±133.0	—
	Wind (W)	Ext. Neg.	—	±37.0	—	—	—	—
		Positive	—	—	—	—	—	—
		Int. Neg.	—	±38.0	—	—	±39.0	—
No.	**Load Combination**							
1	$1.4D + 1.7L$	Ext. Neg.	-7.3	-43.6	-7.3	—	—	—
		Positive	21.8	123.5	61.9	10.7	75.0	39.3
		Int. Neg.	-29.1	-149.1	-76.5	-21.4	-139.4	-71.5
2	$0.75(1.4D + 1.7L + 1.7W)$	Ext. Neg.	-5.4	14.4	-5.4	—	—	—
				-79.9				
		Positive	16.3	92.6	46.4	8.0	56.3	29.5
		Int. Neg.	-21.9	-160.3	-57.4	-16.1	-54.8	-53.6
				-63.4			-154.2	
3	$0.9D + 1.3W$	Ext. Neg.	-3.2	28.6	-3.2	—	—	—
				-67.6				
		Positive	9.7	55.3	27.6	4.8	33.6	17.6
		Int. Neg.	-13.1	-116.1	-34.2	-9.6	-11.7	-32.0
				-17.3			-113.1	
4	$1.54D + 0.55L + 1.43E_h$	Ext. Neg.	-6.3	143.9	-6.3	—	—	—
				-219.3				
		Positive	18.8	106.7	53.4	9.2	64.8	34.0
		Int. Neg.	-25.2	-313.3	-66.1	-18.6	69.8	-61.7
				55.7			-310.6	
5	$0.77D + 1.43E_h$	Ext. Neg.	-2.8	164.9	-2.8	—	—	—
				-198.3				
		Positive	8.3	47.3	23.6	4.1	28.7	15.1
		Int. Neg.	-11.2	-241.5	-29.3	-8.2	136.8	-27.3
				127.4			-243.6	

Design for Flexure – Slab.

The required reinforcement for the slab along column line 6 at the sixth floor level is summarized in Table 6-17. Since $\alpha_1 \ell_2 / \ell_1 > 1.0$, the beams carry the entire gravity load shear forces (see 1913.6.8). Reinforcement details for the slab are shown in Fig. 6-6.

Table 6-17 Required Slab Reinforcement at Floor Level 6

Location			M_u (ft-kips)	b (in.)	d (in.)	A_s (in.^{2})	$A_{s,min}$* (in.^{2})	Reinforcement**
End Span	Column Strip	Ext. Neg.	-7.3	132	6.75	0.24	1.90	10-No. 4
		Positive	21.8	132	6.75	0.72	1.90	10-No. 4
		Int. Neg.	-29.1	132	6.75	0.97	1.90	10-No. 4
	Middle Strip	Ext. Neg.	- 7.3	180	6.75	0.24	2.59	13-No. 4
		Positive	61.9	180	6.75	2.07	2.59	13-No. 4
		Int. Neg.	-76.5	180	6.75	2.57	2.59	13-No. 4
Interior Span	Column Strip	Positive	10.7	132	6.75	0.35	1.90	10-No. 4
		Negative	-21.4	132	6.75	0.71	1.90	10-No. 4
	Middle Strip	Positive	39.3	180	6.75	1.31	2.59	13-No. 4
		Negative	-71.5	180	6.75	2.40	2.59	13-No. 4

*$A_{s,min} = 0.0018bh$ (1913.3.1)

**$s_{max} = 2h = 16$ in. (1913.3.2)

For b = 132 in., 132/16 = 8.3 spaces, say 10 bars. For b = 180 in., 180/16 = 11.3 spaces, say 13 bars.

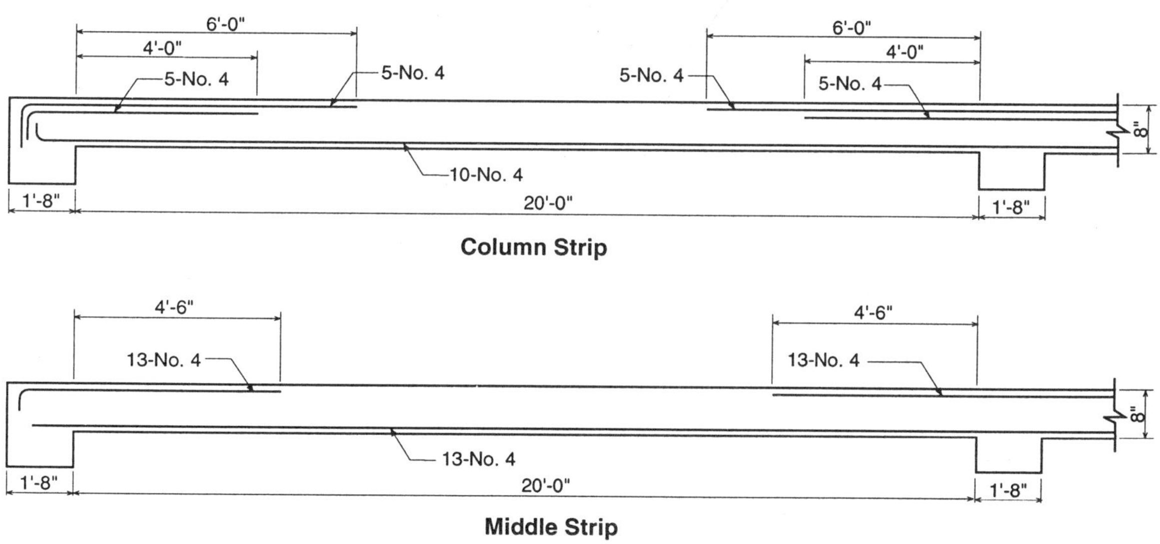

Figure 6-6 Reinforcement Details for Slab Along Column Line 6 on Floor Level 6

Design for Flexure – Beams.

According to 1921.2.1.4, in Seismic Zones 3 and 4, all reinforced concrete members that are part of the lateral load resisting system must satisfy the requirements of 1921.2 through 1921.7, in addition to the requirements of 1901 through 1917. In particular, flexural frame members must satisfy the requirements of 1921.3. A check of the conditions of 1921.3.1 is shown below:

- Factored axial compressive force on beams is negligible. O.K.
- Minimum clear span = 19.83 ft > $4d = 4 \times (24 - 2.5)/12 = 7.2$ ft O.K.
- $\frac{\text{Width}}{\text{Depth}} = \frac{20}{24} = 0.83 > 0.3$ O.K.
- Width = 20 in. > 10 in. O.K.
- Width = 20 in. < width of supporting column + (1.5 × depth of beam)

  < 22 + (1.5 × 24) = 58 in. O.K.

The required reinforcement for the beams on the sixth floor level is shown in Table 6-18. The provided areas of steel are within the limits specified in 1921.3.2.1. Also given in Table 6-18 are the flexural moment strengths (ϕM_n) at each section. According to 1921.3.2.2, the positive moment strength at a joint face must be at least equal to 50% of the negative moment strength provided at that joint. At the exterior negative location, this provision is satisfied since the positive moment strength of 164.1 ft-kips is greater than 220.7/2 = 110.4 ft-kips. The provision is also satisfied at the interior negative location since 164.1 ft-kips is greater than 322.5/2 = 161.3 ft-kips. Additionally, neither the negative nor the positive moment strength at any section along the length of the member shall be less than 25% of the maximum moment strength provided at the face of either joint (1921.3.2.2). In this case, 25% of the maximum moment strength is equal to 322.5/4 = 80.6 ft-kips. Providing at least 2-No. 7 bars (ϕM_n = 113.2 ft-kips) or 2-No. 6 bars (ϕM_n = 83.6 ft-kips) at any section will satisfy this requirement. However, to satisfy the minimum reinforcement requirement of 1921.3.2.1 (i.e., minimum A_s = 1.43 in.^{2}), a minimum of 3-No. 7 top bars (A_s = 1.80 in.^{2}) and 4-No. 6 bottom bars (A_s = 1.76 in.^{2}) must be provided at any section. This also automatically satisfies the requirement that 2 bars be continuous at both the top and bottom of any section (1921.3.2.1).

When reinforcing bars extend through a joint, the column dimension must be at least 20 times the diameter of the largest longitudinal bar for normal weight concrete (1921.5.1.4). In this case, the minimum required column dimension is 20 × 0.875 = 17.5 in. which is less than each of the two column widths that are provided.

Table 6-18 Required Reinforcement for Beams at Floor Level 6

Location		M_u (ft-kips)	A_s* (in.^{2})	Reinforcement*	ϕM_n (ft-kips)
End Span	Ext. Neg.	-219.3	2.38	4-No. 7	220.7
		164.9	1.76	4-No. 6	164.1
	Positive	123.5	1.43	4-No. 6	164.1
	Int. Neg.	-313.3	3.49	6-No. 7	322.5
		127.4	1.43	4-No. 6	164.1
Interior Span	Positive	75.0	1.43	4-No. 6	164.1
	Negative	-310.6	3.46	6-No. 7	322.5
		136.8	1.46	4-No. 6	164.1

**Max. $A_s = 0.025 \times 20 \times 21.5 = 10.75$ in.^{2} (1921.3.2.1)*

Min. $A_s = 3\sqrt{4{,}000} \times 20 \times 21.5/60{,}000 = 1.36$ in.^{2}

$= 200 \times 20 \times 21.5/60{,}000 = 1.43$ in.^{2} (governs)

Design for Shear – Beams.

The beams must be designed for shear forces in accordance with 1921.3.4. These shear forces are determined assuming that moments of opposite sign corresponding to the probable moment strength (M_{pr}) act at the joint faces and that the member is loaded with the tributary gravity load along its span. It is important to note that sidesway to the right and to the left must both be considered when calculating the maximum design shear forces.

The probable moment strength (M_{pr}) is determined for a section using the stress in the tensile reinforcement equal to $1.25f_y$ and a strength reduction factor (ϕ) equal to 1.0. The following equation can be used to compute M_{pr} :

$$M_{pr} = A_s(1.25f_y)\left(d - \frac{a}{2}\right)$$

$$\text{where} \quad a = \frac{A_s(1.25f_y)}{0.85f_c'b}$$

For example, for sidesway to the right, the interior joint must be subjected to the negative moment M_{pr} which is determined as follows:

For 6-No. 7 top bars, $A_s = 6 \times 0.60 = 3.60 \text{ in.}^2$

$$a = \frac{A_s(1.25f_y)}{0.85f_c'b} = \frac{3.60 \times 1.25 \times 60}{0.85 \times 4 \times 20} = 3.97 \text{ in.}$$

$$M_{pr} = A_s(1.25f_y)\left(d - \frac{a}{2}\right) = 3.60 \times 1.25 \times 60 \times \left(21.5 - \frac{3.97}{2}\right) = 5{,}269 \text{ in.-kips} = 439 \text{ ft - kips}$$

Similarly, for the exterior joint, the positive moment M_{pr} based on the 4-No. 6 bottom bars is equal to 225.8 ft-kips. The probable moment strengths for sidesway to the left can be obtained in a similar fashion.

The gravity load on a beam along column line 6 is:

$$w = \frac{8}{12}(145) + 30 + \frac{(16 \times 20) \times 145}{26 \times 144} + 50 = 189 \text{ psf}$$

For beams with $\alpha_1 \ell_2 / \ell_1 \geq 1.0$, the tributary area for shear is bounded by 45-deg lines drawn from the corners of the panel and the centerlines of the adjacent panels parallel to the long sides (1913.6.8.1). For the beams in the direction of analysis, the tributary area is triangular, as depicted in Fig. 6-7. At midspan, the magnitude of the gravity load is equal to $0.189 \times 22 = 4.16$ kips/ft.

Figure 6-8 shows the exterior beam and the shear forces due to the gravity loads. Also shown are the probable moment strengths (M_{pr}) at the joint faces for sidesway to the right and to the left and the corresponding shear forces due to these moments. Note that the maximum combined design shear forces are larger than those obtained from the structural analysis.

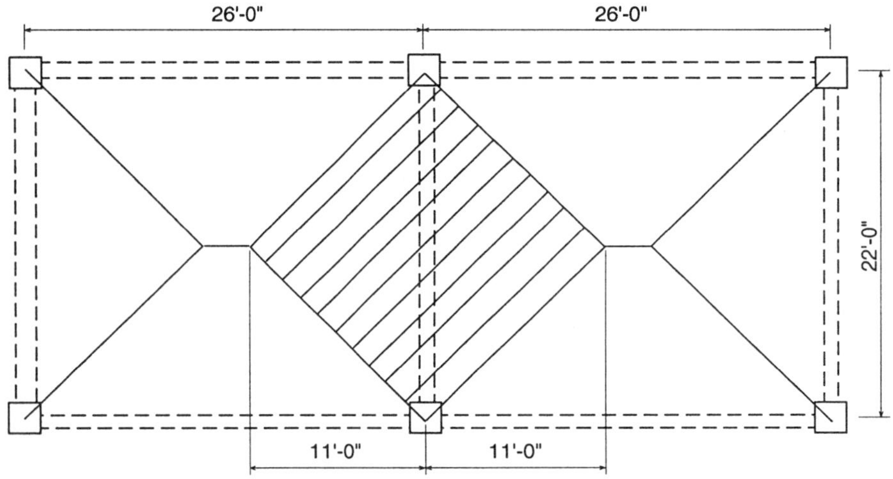

Figure 6-7 Tributary Area for Shear on an Interior Beam

In general, the shear strength is assumed to be provided by both concrete (V_c) and reinforcing steel in the form of hoops or stirrups (V_s). However, the shear strength of the concrete (V_c) is to be taken as zero when the earthquake-induced shear force calculated in accordance with 1921.3.4.1 is greater than or equal to 50% of the total shear force and the factored axial compressive force is less than $A_g f'_c /20$ where A_g is the gross area of the beam (1921.3.4.2). In this particular example, the beam carries negligible axial forces, and the maximum earthquake-induced shear force, which is equal to 33.2 kips (see Fig. 6-8), is more than one-half the total design shear force which is equal to 0.5 × 56.1 = 28.1 kips. Thus, V_c must be zero. The maximum shear force V_s is (1911.1):

$$\phi V_s = V_u - \phi V_c$$

$$V_s = \frac{V_u}{\phi} - 0 = \frac{56.1}{0.85} = 66.0 \text{ kips}$$

where the strength reduction factor (ϕ) was taken as 0.85 since the shear force (V_u) corresponds to the nominal flexural strength of the member (see 1909.3.4.1).

Shear strength contributed by shear reinforcement shall not exceed $(V_s)_{max}$ (1911.5.6.8):

$$(V_s)_{max} = 8\sqrt{f_c'}\, b_w d = 8 \cdot \sqrt{4{,}000} \times 20 \times 21.5 / 1{,}000 = 217.6 \text{ kips} > 66.0 \text{ kips} \quad \text{O.K.}$$

Also, V_s is less than $4\sqrt{f_c'} b_w d$ = 108.8 kips.

Required spacing of No. 4 closed stirrups (hoops) for a factored shear force of 66.0 kips is (1911.5.6.2):

$$s = \frac{A_v f_y d}{V_s} = \frac{(4 \times 0.2) \times 60 \times 21.5}{66.0} = 15.6 \text{ in.}$$

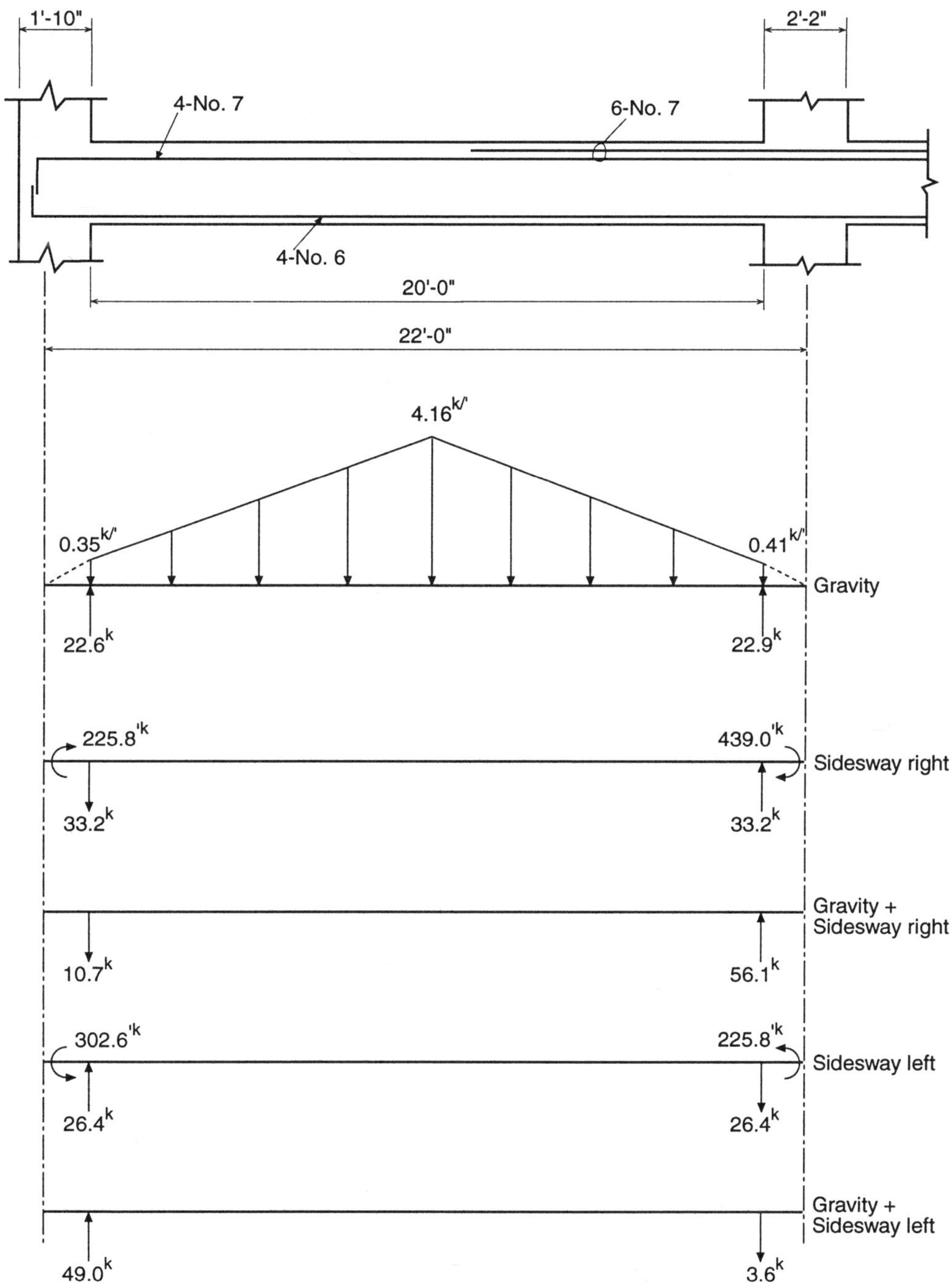

Figure 6-8 Design Shear Forces for Exterior Beam Along Column Line 6 on Floor Level 6

Note that 4 legs are required in order to conform with the provisions of 1921.3.3.3 for lateral support of the longitudinal bars.

Maximum allowable hoop spacing (s_{max}) within a distance of $2h = 2 \times 24 = 48$ in. (1921.3.3.1) from the face of the support is the smallest of the following (1921.3.3.2):

$$s_{max} = \frac{d}{4} = \frac{21.5}{4} = 5.4 \text{ in. (governs)}$$

$$= 8 \times (\text{diameter of smallest longitudinal bar}) = 8 \times 0.75 = 6.0 \text{ in.}$$

$$= 24 \times (\text{diameter of hoop bar}) = 24 \times 0.50 = 12.0 \text{ in.}$$

$$= 12 \text{ in.}$$

Therefore, hoops must be spaced at 5 in. on center with the first one located at 2 in. from the face of the support. Eleven hoops are to be placed at this spacing. Where hoops are no longer required, stirrups with seismic hooks at both ends may be used (1921.3.3.4). At a distance of 52 in. from the face of the interior support, $V_u = 50.8$ kips. Also, with $V_c = 54.4$ kips, the spacing required for No. 4 sitrrups with two legs is 23.5 in.; this spacing is greater than the maximum allowable spacing which is $d/2 = 10.8$ in. (1921.3.3.4). A 10 in. spacing, starting at 52 in. from the face of the support will be sufficient for the remaining portion of the beam.

Reinforcing Bar Cutoff Points – Beams.

The negative reinforcement at the interior support is 6-No. 7 bars. The location where three of the six bars can be terminated will be determined. Note that 3-No. 7 bars must be continuous throughout the length of the beam to satisfy the minimum reinforcement requirements of 1921.3.2.1.

The loading used to find the cutoff point of the 3-No. 7 bars at this location is 0.9 times the dead load in combination with the probable flexural strengths (M_{pr}) at the ends of the member; this combination will produce the longest bar lengths. The design flexural strength (ϕM_n) provided by 3-No. 7 bars is 167.7 ft-kips. Therefore, the 3 reinforcing bars can be terminated after the required moment strength (M_u) has been reduced to 167.7 ft-kips.

With $\phi = 1.0$ and $f_s = 75$ ksi, M_{pr} is equal to 439.0 ft-kips at the interior support and 225.8 ft-kips at the exterior support. The load is equal to $0.9 \times (139 \times 22)/1{,}000 = 2.75$ kips/ft at midspan. The distance from the face of the interior support to where the moment equals 167.7 ft-kips is obtained by summing moments about section *a-a* in Fig. 6-9:

$$\frac{x}{2}\left[\frac{(2.75-0.27)x}{9.9167}\right]\left(\frac{x}{3}\right)+\frac{0.27x^2}{2}-48.3x+439.0=167.7$$

Solving for x gives a distance of 5.9 ft. The 3-No. 7 bars must extend a distance $d = 21.5$ in. or $12d_b = 12 \times 0.875 = 10.5$ in. beyond the distance x (1912.10.3). Thus, from the face of the support, the total bar length must be at least equal to $5.9 + (21.5/12) = 7.7$ ft. Also, the bars must extend a full development length ℓ_d beyond the face of the support (1912.10.4). The development length can be determined from the provisions of 1912.2.3:

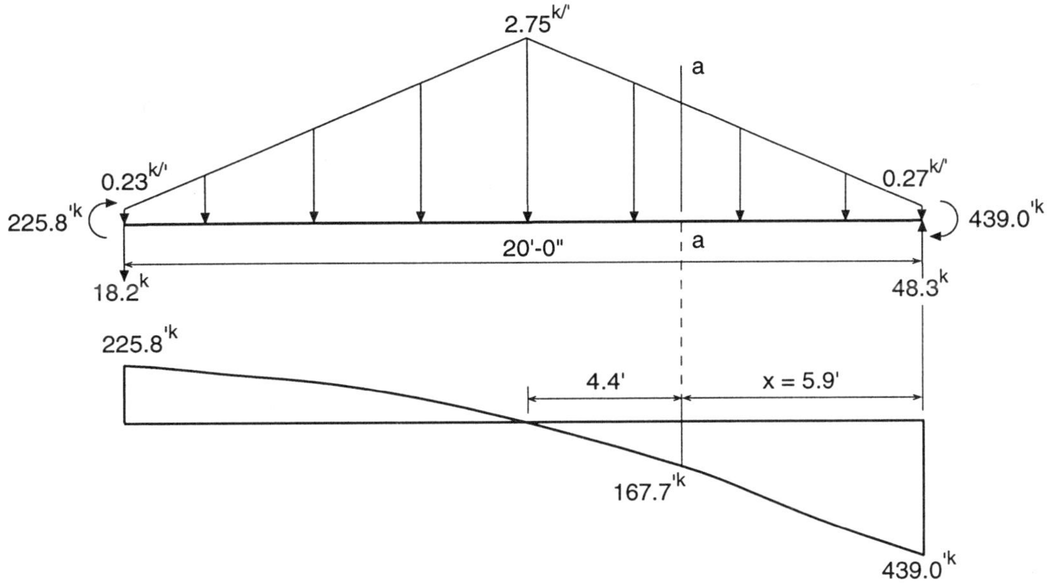

Figure 6-9 Moment Diagram for Cutoff Location of Negative Bars at Interior Support

$$\frac{\ell_d}{d_b} = \frac{3}{40}\frac{f_y}{\sqrt{f_c'}}\frac{\alpha\beta\gamma\lambda}{\left(\dfrac{c+K_{tr}}{d_b}\right)}$$

where α = reinforcement location factor = 1.3 (top bar)

β = coating factor = 1.0 (uncoated reinforcement)

γ = reinforcement size factor = 1.0 (No. 7 bar)

λ = lightweight aggregate concrete factor = 1.0 (normal weight concrete)

$$c = \text{spacing or cover dimension} = \begin{cases} 1.5 + 0.5 + \dfrac{0.875}{2} = 2.44 \text{ in.} \\ \dfrac{20 - 2(1.5+0.5) - 0.875}{2 \times 5} = 1.5 \text{ in. (governs)} \end{cases}$$

K_{tr} = transverse reinforcement index = 0 (conservative)

$$\frac{c+K_{tr}}{d_b} = \frac{1.5+0}{0.875} = 1.7$$

$$\frac{\ell_d}{d_b} = \frac{3}{40} \times \frac{60{,}000}{\sqrt{4{,}000}} \times \frac{1.3 \times 1.0 \times 1.0 \times 1.0}{1.7} = 54.4$$

$$\ell_d = 54.4 \times 0.875 = 47.6 \text{ in.} = 4.0 \text{ ft} < 7.7 \text{ ft}$$

Thus, the total required length of the 3-No. 7 bars must be at least 7.7 ft beyond the face of the support. It is important to note that flexural reinforcement shall not be terminated in a tension zone unless one or more of the conditions of 1912.10.5 are satisfied. In this case, the point of inflection is approximately 10.3 ft from the face of the right support which is greater than 7.7 ft. Therefore, the 3-No. 7 bars can not be terminated here unless one of the conditions of 1912.10.5 is satisfied. Using 1912.10.5.1, check if the factored shear force (V_u) at the cutoff point does not exceed two-thirds of that permitted ($2\phi V_n/3$). For No. 4 stirrups at 10 in. on center that are provided in this region of the beam:

$$\phi V_n = \phi(V_s + V_c) = 0.85 \times \left(\frac{0.4 \times 60 \times 21.5}{10} + 54.4\right) = 90.1 \text{ kips}$$

$$\frac{2}{3}\phi V_n = 60.1 \text{ kips} > V_u = 38.8 \text{ kips at 7.7 ft from face of support}$$

Since $2\phi V_n/3 > V_u$, the cutoff point for the 3-No. 7 bars can be 7.7 ft beyond the face of the interior support.

The cutoff point for one of the 4-No. 7 bars at the exterior support can be determined in a similar fashion. In order to satisfy the minimum reinforcement requirements of 1921.3.2.1, the 4-No. 6 bottom bars must be continuous. Note that the reinforcing bars that terminate at exterior columns must be properly developed by providing 90-deg hooks embedded in the confined core of the column (1921.5.1.3). The development length of a hook (ℓ_{dh}) is the largest of the following (1921.5.4.1):

$$\ell_{dh} = 8d_b = 8 \times 0.875 = 7.0 \text{ in.}$$

$$= 6 \text{ in.}$$

$$= \frac{f_y d_b}{65\sqrt{f'_c}} = \frac{60{,}000 \times 0.875}{65\sqrt{4{,}000}} = 12.8 \text{ in.} \quad \text{(governs)}$$

Therefore, the hooks must extend at least 13 in. into the column with a $12d_b = 10.5$ in. extension (1907.1.2). Figure 6-10 shows the reinforcement details for the exterior beam.

6.6.3 Column Design

This section outlines the design of an interior column and an edge column along column line 6 below the first floor level. A summary of the design forces is given in Tables 6-19 and 6-20 for column C6 and Tables 6-21 and 6-22 for column D6. Note that in Seismic Zones 2, 3 and 4, the orthogonal effects of the earthquake forces must be considered unless the axial load in the column due to seismic forces in either direction is less than 20% of the column axial load capacity (1633.1). Conservatively assuming a 1% percent reinforcement ratio for the 26 × 26 in. column, the axial load capacity ($\phi P_{n(max)}$) is equal to 1,501 kips (1910.3.5.1); 20% of this load is 300 kips which is greater than the axial load of 51 kips due to seismic forces in the N-S direction (see Fig. 6-2) and 15 kips due to seismic forces in the E-W direction. Similarly, for the 22 × 22 in. column, $0.2\phi P_{n(max)}$ is equal to 215 kips which is greater than the axial load of 125 kips and 2 kips due to seismic forces in the N-S and E-W directions, respectively. Thus, orthogonal effects need not be considered.

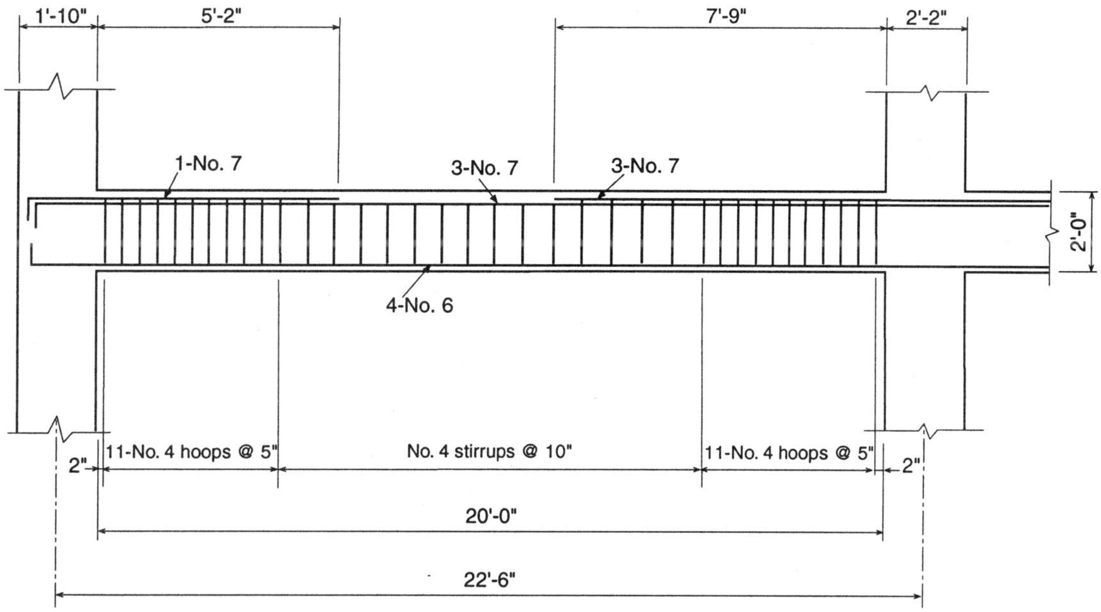

Figure 6-10 Reinforcement Details for Exterior Beam Along Column Line 6 on Floor Level 6

The requirements of 1921.4 apply to members resisting earthquake forces and factored axial forces exceeding $A_g f'_c/10$. For a 26 × 26 in. column with f'_c = 4 ksi, $A_g f'_c/10$ = 270 kips which is less than the maximum factored axial load on the column (see Table 6-20). Thus, the following two criteria must be satisfied:

- Shortest cross-sectional dimension = 26 in. > 12 in.   O.K.
- Ratio of shortest cross-sectional dimension to perpendicular dimension = 1.0 > 0.4   O.K.

These requirements are also satisfied for the 22 × 22 in. column.

Table 6-19 Summary of Factored Axial Loads and Bending Moments for Column C6 between the 1st and 2nd Floor Levels

	Load Case	Axial Load (kips)	Bending Moment (ft-kips)
	Dead (*D*)	1,024	1.0†
	Live (*L*)*	122	10.4†
	Earthquake (E_h)	±49	±119.0
	Wind (*W*)	±14	±32.0
No.	Load Combination		
1	$1.4D + 1.7L$	1,641	19.1
2	$0.75(1.4D + 1.7L + 1.7W)$	1,213	55.1
		1,249	-26.5
3	$0.9D + 1.3W$	903	42.5
		940	-40.7
4	$1.54D + 0.55L + 1.43E_h$	1,574	177.4
		1,714	-162.9
5	$0.77D + 1.43E_h$	718	170.9
		859	-169.4

**Includes live load reduction (see Table 6-12)*
† *From Eq. (13-4)*

Table 6-20 Summary of Factored Axial Loads and Bending Moments for Column C6 between the Ground and 1st Floor Levels

No.	Load Case	Axial Load (kips)	Bending Moment (ft-kips)
	Dead (D)	1,121	1.0[†]
	Live (L)*	133	7.8[†]
	Earthquake (E_h)	±51	±172.0
	Wind (W)	±14	±34.0
No.	**Load Combination**		
1	$1.4D + 1.7L$	1,796	14.7
2	$0.75(1.4D + 1.7L + 1.7W)$	1,329	54.3
		1,365	-32.4
3	$0.9D + 1.3W$	991	45.1
		1,027	-43.3
4	$1.54D + 0.55L + 1.43E_h$	1,727	251.8
		1,872	-240.1
5	$0.77D + 1.43E_h$	790	246.7
		936	-245.2

**Includes live load reduction (see Table 6-12)*

† From Eq. (13-4)

Table 6-21 Summary of Factored Axial Loads and Bending Moments for Column D6 between the 1st and 2nd Floor Levels

No.	Load Case	Axial Load (kips)	Bending Moment (ft-kips)
	Dead (D)	642	-14.5[†]
	Live (L)*	71	-5.2[†]
	Earthquake (E_h)	±119	±58.0
	Wind (W)	±34	±16.0
No.	**Load Combination**		
1	$1.4D + 1.7L$	1,020	-29.1
2	$0.75(1.4D + 1.7L + 1.7W)$	721	-1.5
		808	-42.3
3	$0.9D + 1.3W$	534	7.8
		622	-33.9
4	$1.54D + 0.55L + 1.43E_h$	858	57.8
		1,198	-108.1
5	$0.77D + 1.43E_h$	324	71.8
		665	-94.1

**Includes live load reduction (see Table 6-13)*

† Design strip moment/2 (see Table 6-11)

Table 6-22 Summary of Factored Axial Loads and Bending Moments for Column D6 between the Ground and 1st Floor Levels

No.	Load Case	Axial Load (kips)	Bending Moment (ft-kips)
	Dead (D)	703	-12.4[†]
	Live (L)*	77	-4.5[†]
	Earthquake (E_h)	±125	±88.0
	Wind (W)	±36	±17.0
No.	Load Combination		
1	$1.4D + 1.7L$	1,115	-25.0
2	$0.75(1.4D + 1.7L + 1.7W)$	790	2.9
		882	-40.4
3	$0.9D + 1.3W$	586	10.9
		680	-33.3
4	$1.54D + 0.55L + 1.43E_h$	946	104.3
		1,304	-147.4
5	$0.77D + 1.43E_h$	363	116.3
		720	-135.4

**Includes live load reduction (see Table 6-13)*
[†] Design strip moment/2 (see Table 6-11)

Based on the load combinations in Table 6-20, a 26 × 26 in. column with 12-No. 11 bars (ρ_g = 2.77%) is adequate for column C6 below the first floor level. Figure 6-11, obtained from *PCACOL* (Ref. 6.3), shows the interaction diagram for this column. Also, 12-No. 9 bars (ρ_g = 2.48%) is adequate for column D6 below the first floor level (see Fig. 6-12). Note that the provided reinforcement ratios for both columns are within the allowable range of 1% and 6% (1921.4.3.1).

Relative Flexural Strengths of Columns and Girders.

Section 1921.4.2 requires that the sum of the flexural strengths of the columns at a joint must be greater than or equal to 6/5 the sum of the flexural strengths of the girders framing into that joint. Note that the column flexural strength is to be determined for the factored axial force resulting in the lowest flexural strength, consistent with the direction of the lateral forces considered. Only the seismic load combinations need to be considered when checking the relative strengths of the columns and girders.

At interior column C6, the negative design flexural strength of the beam is 322.5 ft-kips (see Table 6-18). The positive design flexural strength of the beam on the opposite face of the column is 164.1 ft-kips. For the upper column framing into the joint, the minimum flexural strength (ϕM_n) is approximately 495.0 ft-kips, which corresponds to a factored axial load of 1,714 kips (see Table 6-19 and pt. 7 in Fig. 6-13). Similarly, the minimum flexural strength is about 385.1 ft-kips for the lower column framing into the joint; this corresponds to a factored axial load of 1,872 kips (see pt. 7 in Fig. 6-11). Therefore,

$$\Sigma M_e = 495.0 + 385.1 = 880.1 \text{ ft-kips}$$

$$\Sigma M_g = 322.5 + 164.1 = 486.6 \text{ ft-kips}$$

$$880.1 \text{ ft-kips} > \frac{6}{5} \times 486.6 = 583.9 \text{ ft-kips} \quad \text{O.K.}$$

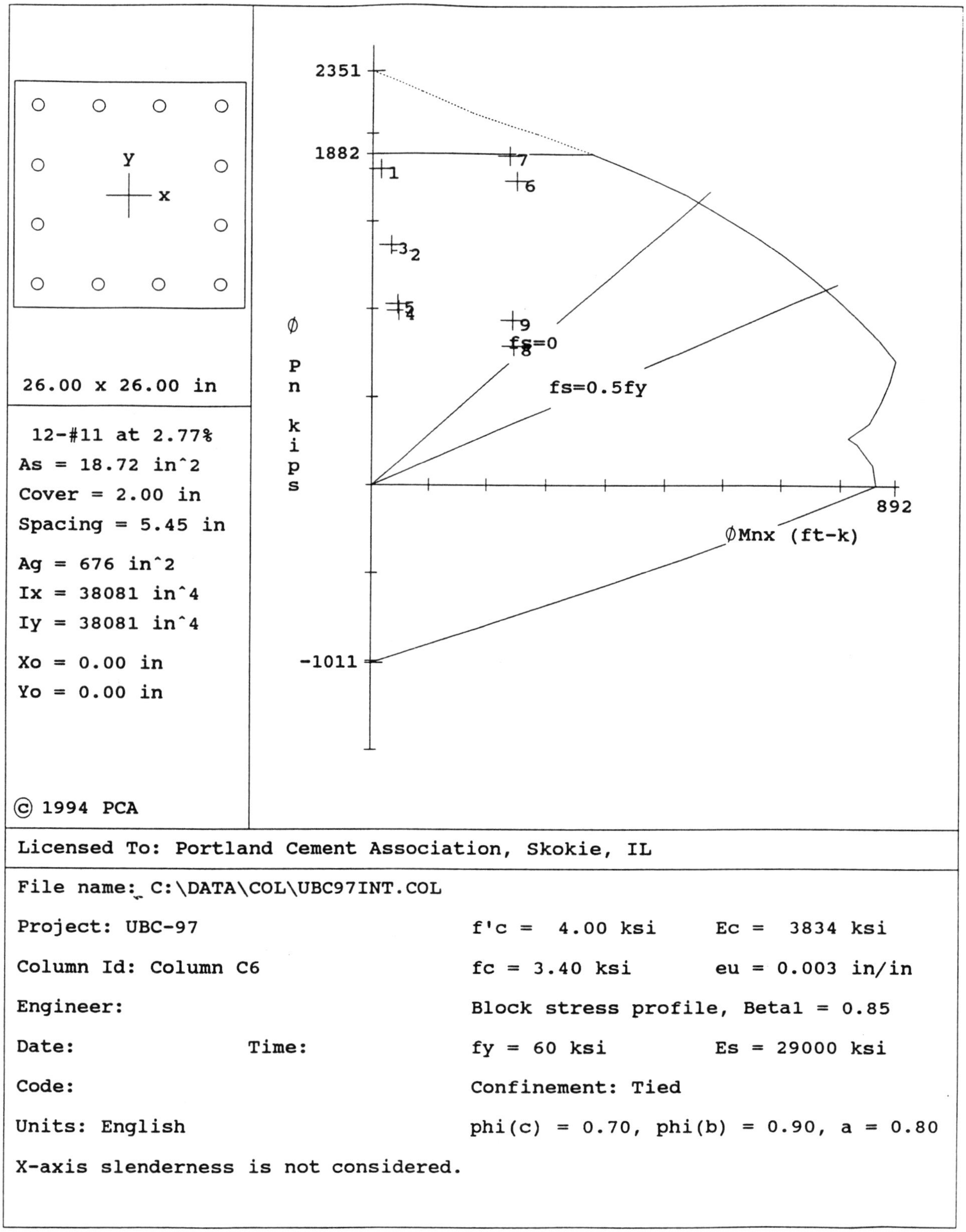

Figure 6-11 Design Strength Interaction Diagram for Column C6 Below the First Floor Level

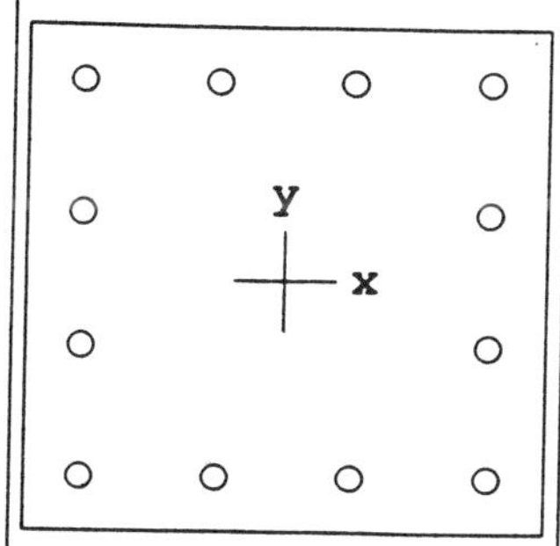

22.00 x 22.00 in

12-#9 at 2.48%
As = 12.00 in^2
Cover = 1.88 in
Spacing = 4.58 in

Ag = 484 in^2
Ix = 19521 in^4
Iy = 19521 in^4

Xo = 0.00 in
Yo = 0.00 in

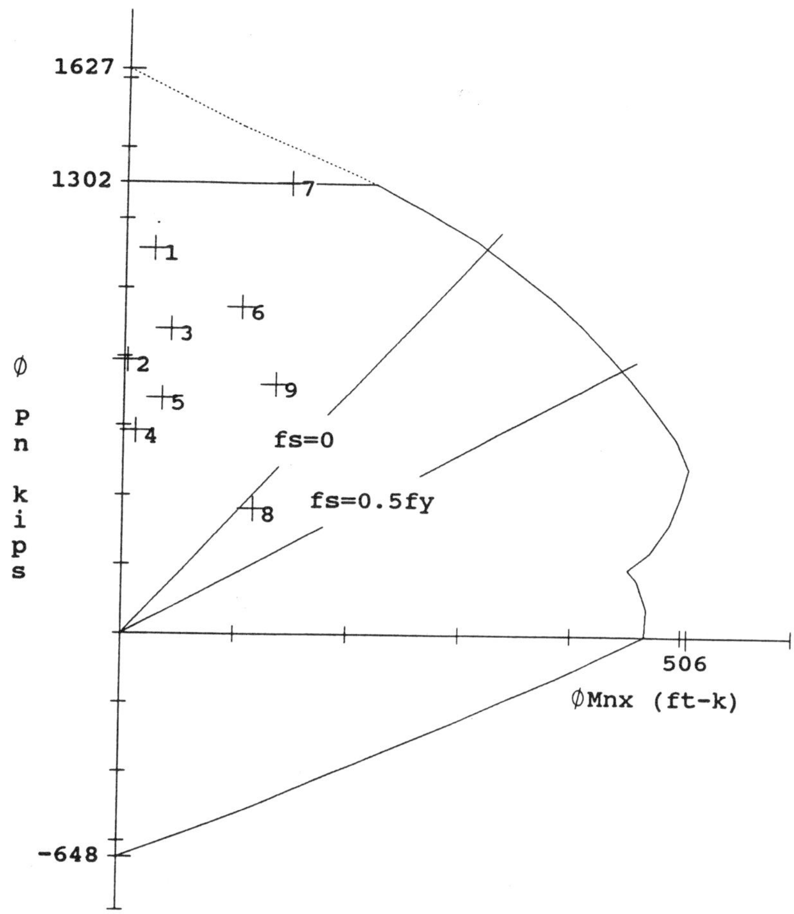

File name: C:\DATA\COL\UBC97EXT.COL

Project: UBC-97		f'c = 4.00 ksi	Ec = 3834 ksi
Column Id: Column D6		fc = 3.40 ksi	eu = 0.003 in/in
Engineer:		Block stress profile, Beta1 = 0.85	
Date:	Time:	fy = 60 ksi	Es = 29000 ksi
Code:		Confinement: Tied	
Units: English		phi(c) = 0.70, phi(b) = 0.90, a = 0.80	

X-axis slenderness is not considered.

Figure 6-12 Design Strength Interaction Diagram for Column D6 Below the First Floor Level

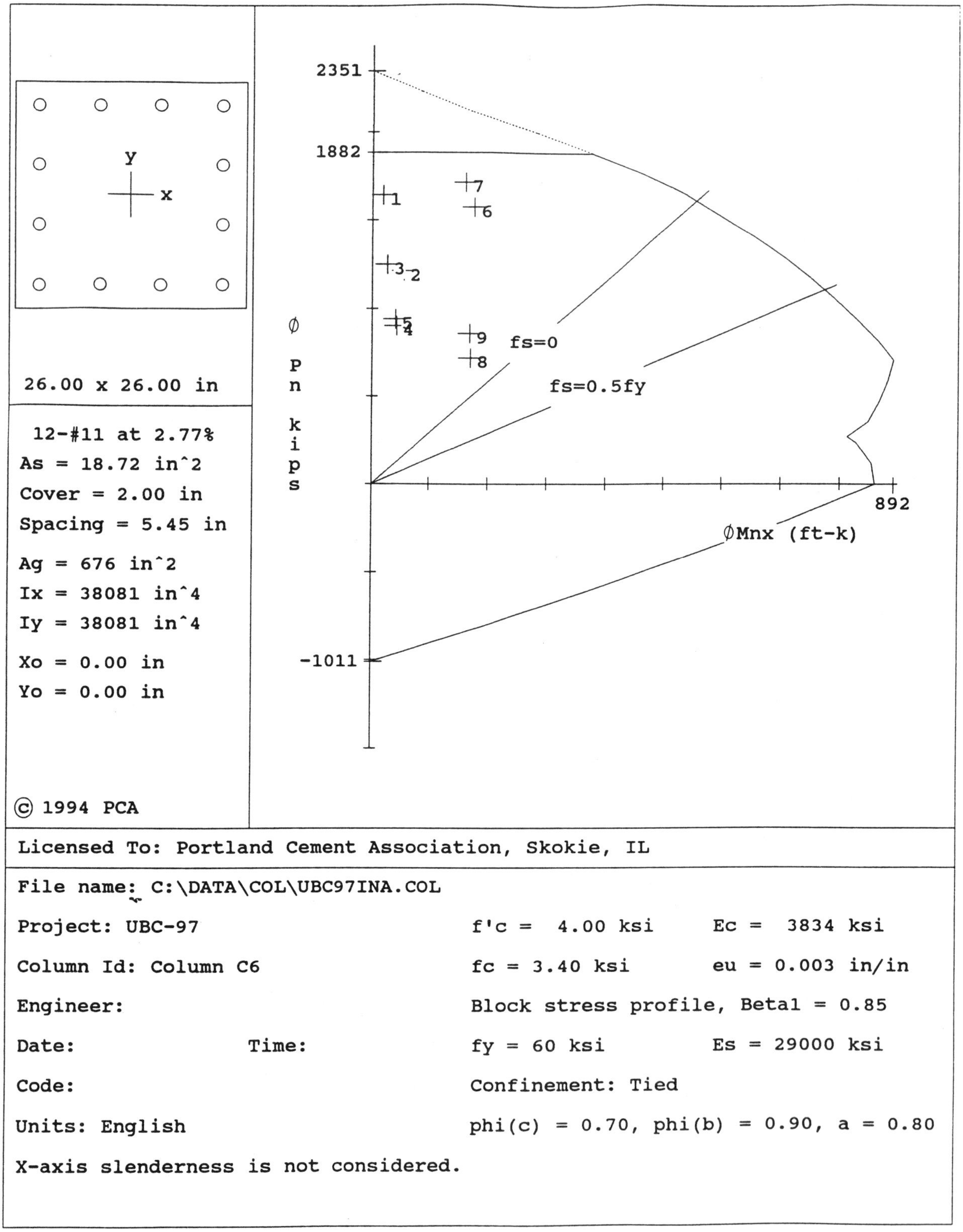

Figure 6-13 Design Strength Interaction Diagram for Column C6 between the First and Second Floor Levels

At exterior column D6, the negative design flexural strength of the beam is 220.7 ft-kips (see Table 6-18). For the upper column framing into the joint, the design flexural strength corresponding to a factored axial load of 1,198 kips is 282.2 ft-kips (see pt. 7 in Fig. 6-14). For the lower column, the design flexural strength corresponding to a factored axial load of 1,304 kips is 219.9 ft-kips (see pt. 7 in Fig. 6-12). Therefore,

$$282.2 + 219.9 = 502.1 \text{ ft-kips} > \frac{6}{5} \times 220.7 = 264.8 \text{ ft-kips} \quad \text{O.K.}$$

Confinement Reinforcement.

Special transverse reinforcement for confinement is required over a distance ℓ_o at the column ends where ℓ_o equals the maximum of (1921.4.4.4):

Depth of member = 22 in. for column D6 and 26 in. for column C6

Clear height/6 = [(16 × 12) - 24]/6 = 28 in. (governs)

18 in.

According to 1921.4.4.2, the maximum allowable spacing of rectangular hoops within the 28 in. length shall not exceed one-fourth of the minimum member dimension which is equal to 0.25 × 26 = 6.5 in. for column C6 and 5.5 in. for column D6, or 4 in. (governs).

The minimum required cross-sectional area of hoop reinforcement (A_{sh}) is the larger value obtained from the following two equations (1921.4.4.1):

$$A_{sh} = \frac{0.3 s h_c f_c'}{f_{yh}} \left[\left(\frac{A_g}{A_{ch}} \right) - 1 \right]$$

$$= \frac{0.09 s h_c f_c'}{f_{yh}}$$

where s = spacing of transverse reinforcement

h_c = cross-sectional dimension of column core, measured center-to center of confining reinforcement

A_g = gross area of section

A_{ch} = cross-sectional area of member measured out-to-out of transverse reinforcement

f_{yh} = specified yield strength of transverse reinforcement

Using a hoop spacing of 4 in., f_{yh} = 60,000 psi, a clear cover of 1.5 in., and tentatively assuming No. 4 hoops, the required cross-sectional area of hoop reinforcement for column C6 is:

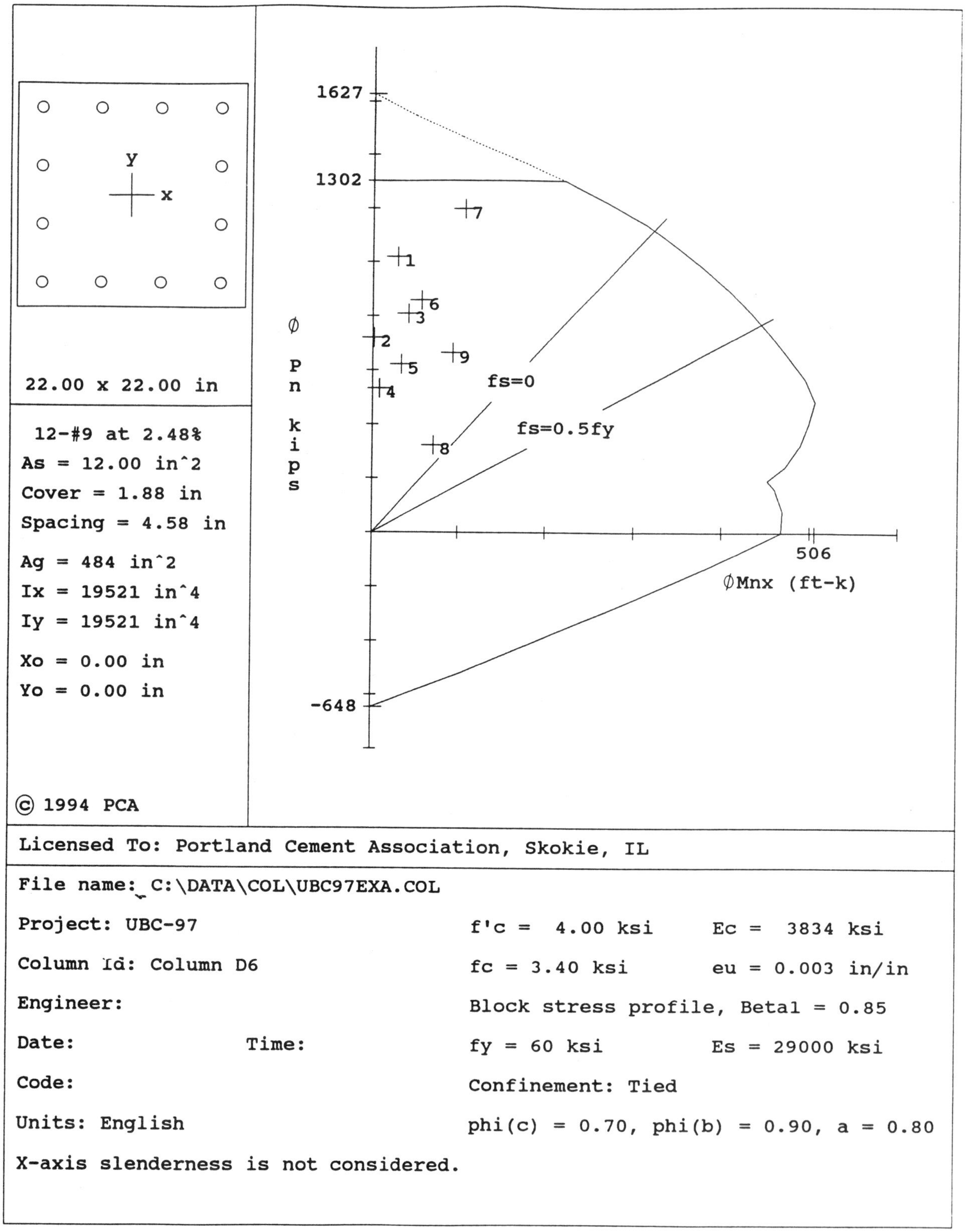

Figure 6-14 Design Strength Interaction Diagram for Column D6 between the First and Second Floor Levels

$$A_{sh} = \frac{0.3 \times 4 \times [26 - (2 \times 1.5) - 0.5] \times 4}{60} \times \left[\frac{26^2}{(26-3)^2} - 1 \right] = 0.50 \text{ in.}^2$$

$$= \frac{0.09 \times 4 \times [26 - (2 \times 1.5) - 0.5] \times 4}{60} = 0.54 \text{ in.}^2 \qquad \text{(governs)}$$

Using No. 4 hoops with two crossties provides $A_{sh} = 4 \times 0.2 = 0.60$ in.^{2} which is greater than 0.54 in.^{2} Use 4 in. spacing for the hoops at the column ends. Similar calculations can be performed for column D6.

Transverse Reinforcement for Shear.

Similar to beams, the design shear forces for columns shall be determined considering the maximum forces that can be generated at the faces of the joints at each end (1921.4.5.1). These joint forces shall be determined using the probable moment strengths (M_{pr}) associated with the range of factored axial loads on the member; M_{pr} is calculated with the strength reduction factors equal to 1.0 and assuming that the stress in the tensile reinforcement is equal to $1.25f_y$. The member shear forces need not exceed those determined from joint strengths based on the probable moment strength of the transverse members framing into the joint. It is important to note that the calculated shear force must not be taken less than the factored shear force obtained from analysis.

Figure 6-15 is the interaction diagram for column C6 with the strength reduction factors equal to 1.0 and f_y equal to $1.25 \times 60 = 75$ ksi. Conservatively assume that the maximum moment (M_{pr}) is equal to the moment at the balanced point which is approximately 1,374 ft-kips. Thus, the design shear force based on the probable moment strength at the column ends is equal to $(2 \times 1{,}374)/[16 - (24/12)] = 196.3$ kips.

The positive probable moment strength of the beam framing into the joint at the face of column C6 at the first floor level is 225.8 ft-kips (see Fig. 6-8). The negative probable moment strength of the beam on the other side of the column is 439.0 ft-kips. Assuming that the bending moments at the top and bottom of the column supporting the first floor level are equal, the design shear force based on the probable moment strengths of the beams is:

$$V_u = \frac{2 \times (M_{pr}^+ + M_{pr}^-) \times \left(\frac{\ell_2}{\ell_1 + \ell_2} \right)}{\ell_1} = \frac{2 \times (225.8 + 439.0) \times \left(\frac{10}{14 + 10} \right)}{14} = 39.6 \text{ kips}$$

where ℓ_1 and ℓ_2 are the clear lengths of the columns in the first and second stories, respectively. This factored shear force is greater than the factored shear force determined from analysis. Since the factored axial forces for column C6 are greater than $A_g f'_c/20 = (26^2 \times 4)/20 = 135$ kips, the shear strength of the concrete may be used (1921.4.5.2 and 1911.3.1.2):

$$V_c = 2\left(1 + \frac{N_u}{2{,}000\,A_g}\right)\sqrt{f'_c}\,b_w d = 2 \times \left(1 + \frac{790{,}000}{2{,}000 \times 26^2}\right) \times \sqrt{4{,}000} \times 26 \times 23.3 / 1{,}000 = 121.4 \text{ kips}$$

$\phi V_c = 0.85 \times 121.4 = 103.2 \text{ kips} > 39.6 \text{ kips}$

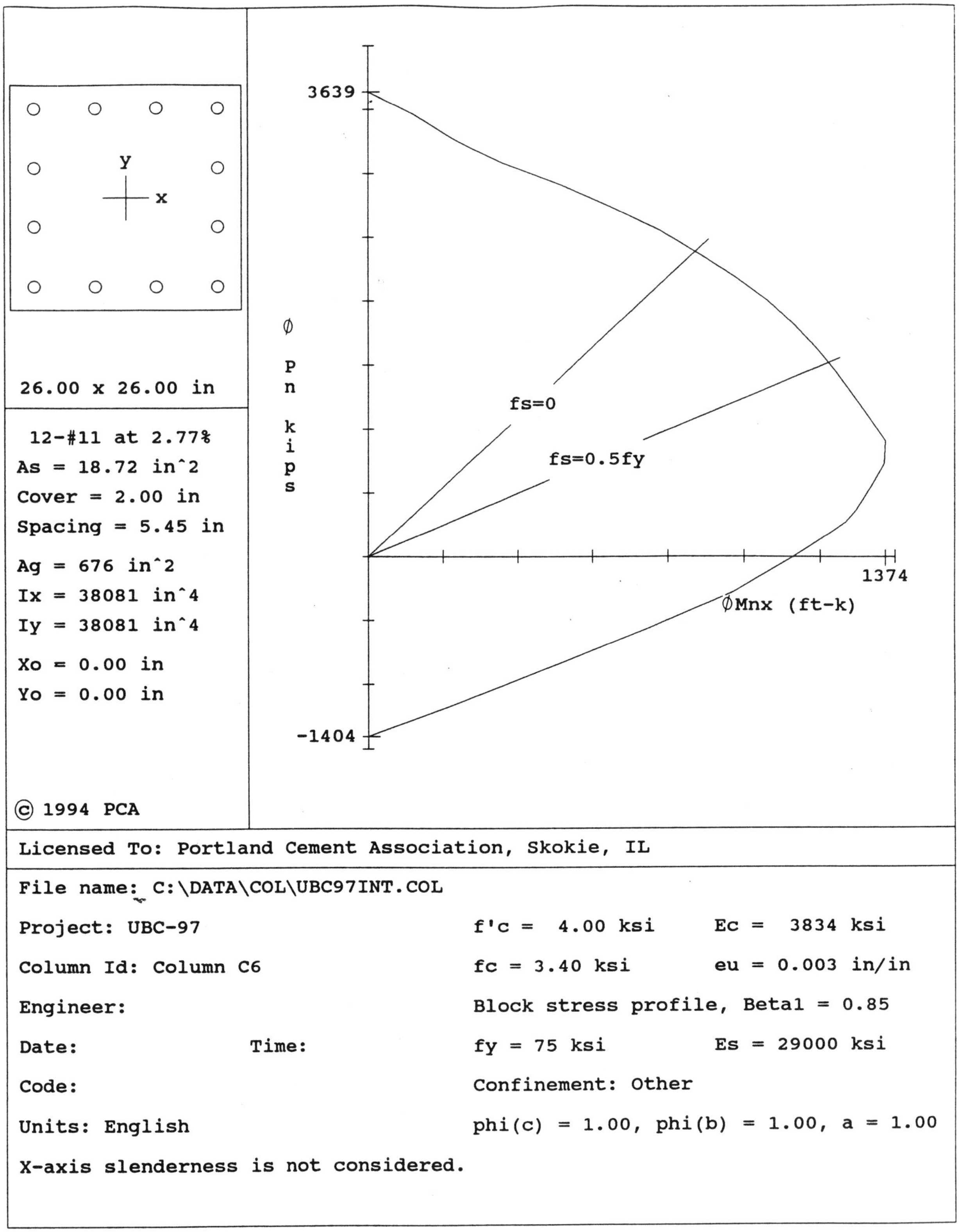

Figure 6-15 Design Strength Interaction Diagram for Column C6 with f_y = 75 ksi and ϕ = 1.0

Thus, the No. 4 hoops at 4 in. on center required for confinement over the distance ℓ_o at the column ends is also adequate for the transverse reinforcement required for shear.

The remainder of the column length must contain hoop reinforcement with center-to-center spacing not to exceed either six times the diameter of the longitudinal column bars which is equal to 6 × 1.41 = 8.5 in. or 6 in. (1921.4.4.6). In this case, the 6 in. spacing controls.

The shear strength requirements for column D6 can be checked in a similar fashion.

Splice Length for Column Vertical Bars.

The lap splice length of the No. 11 bars in column C6 can be determined according to 1912.2.3:

$$\frac{\ell_d}{d_b} = \frac{3}{40} \frac{f_y}{\sqrt{f'_c}} \frac{\alpha\beta\gamma\lambda}{\left(\dfrac{c + K_{tr}}{d_b}\right)}$$

where α = reinforcement location factor = 1.0 (no top bar effect)

β = coating factor = 1.0 (uncoated reinforcement)

γ = reinforcement size factor = 1.0 (No. 7 or larger bar size)

λ = lightweight aggregate concrete factor = 1.0 (normal weight concrete)

$$K_{tr} = \text{transverse reinforcement index} = \frac{A_{tr} f_{yt}}{1{,}500 sn} = \frac{3 \times 0.2 \times 60{,}000}{1{,}500 \times 4 \times 4} = 1.5$$

$$\frac{c + K_{tr}}{d_b} = \frac{2.7 + 1.5}{1.41} = 3.0 > 2.5, \text{ use } 2.5$$

$$\frac{\ell_d}{d_b} = \frac{3}{40} \times \frac{60{,}000}{\sqrt{4{,}000}} \times \frac{1.0 \times 1.0 \times 1.0 \times 1.0}{2.5} = 28.5$$

$$\ell_d = 28.5 \times 1.41 = 40.1 \text{ in.} = 3.3 \text{ ft}$$

According to 1921.4.3.2, the lap splices must be located within the center half of the column length and must be proportioned as tension splices. From Fig. 6-11 it can be seen that the stress in the longitudinal bars is compressive for all load combinations. However, a Class B splice is required since all of the bars will be spliced at the same location (1912.15.2). Therefore, the Class B splice length is equal to 1.3 × 3.3 = 4.3 ft. Reinforcement details for column C6 are shown in Fig. 6-16.

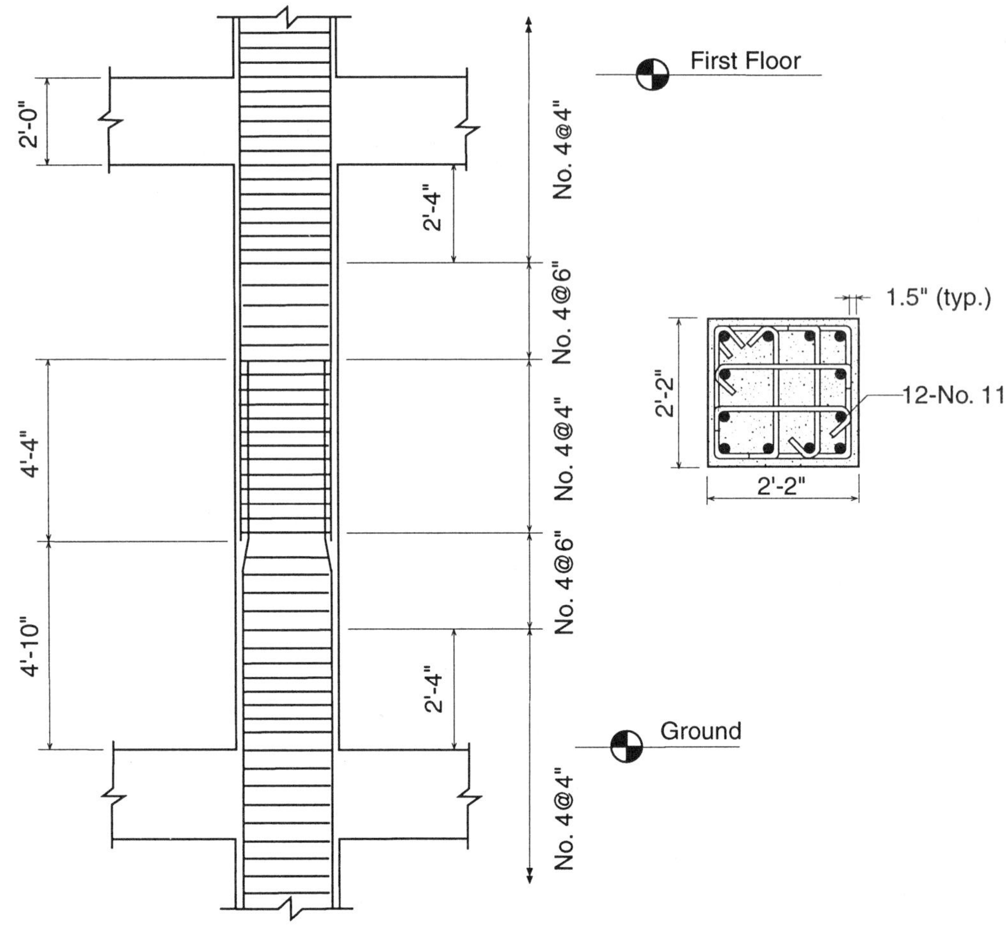

Figure 6-16 Reinforcement Details for Column C6

6.6.4 Design of Beam-Column Joint

Interior Joint.

Section 1921.5 requires that transverse hoop reinforcement as specified in 1921.4.4 shall be provided within the joint, unless the joint is confined by structural members as specified in 1921.5.2.2. For an interior joint, beams frame into all four sides of the joint, and their width is 77% of the column width. Thus, within the joint depth, transverse reinforcement equal to at least one-half the amount required by 1921.4.4.1 shall be provided, and the spacing is permitted to be $0.25 \times 26 = 6.5$ in. or 6 in. (governs). For detailing simplicity, use the transverse reinforcement of the column (No. 4 @ 4 in.) through the joint.

Figure 6-17 shows the interior beam-column joint at the first floor level. The shear strength is checked in the N-S direction along column line 6. The shear force at section x-x is obtained by subtracting the column horizontal shear force from the sum of the tensile force in the top beam reinforcement and the compressive force at the top of the beam on the opposite face of the column.

The column horizontal shear force (V_h) can be obtained by assuming that the adjoining floors are deformed so that plastic hinges form at their junctions with the columns, with M_{pr}^-(beam) = 439.0 ft - kips and M_{pr}^+(beam) = 225.8 ft - kips (1921.5.1.1).

It is assumed that the beam moments are resisted by the columns above and below the joint inversely proportional to their lengths. Thus, above the joint, the moment is equal to (439.0 + 225.8) × [14/(14 + 10)] = 387.8 ft-kips; below the joint the moment is equal to 277.0 ft-kips. The horizontal shear force at the top of the joint is then equal to:

$$V_h \text{ (top)} = \frac{2 \times 387.8}{10} = 77.6 \text{ kips}$$

and at the bottom of the joint it is:

$$V_h \text{ (bottom)} = \frac{2 \times 277.0}{14} = 39.6 \text{ kips}$$

The tensile force in the top reinforcement (6-No. 7 bars) is equal to (6 × 0.60) × 1.25 × 60 = 270.0 kips. The compressive force on the opposite side of the column is equal to the tensile force of 4-No. 6 bars which is (4 × 0.44) × 1.25 × 60 = 132.0 kips.

The net top shear at section *x-x* is $T_1 + C_2 - V_h$ (top) = 270.0 + 132.0 - 77.6 = 324.4 kips.

The net bottom shear at section *x-x* is $T_2 + C_1 - V_h$ (bottom) = 132.0 + 270.0 - 39.6 = 362.4 kips.

For a joint confined on all four faces, the nominal shear strength (ϕV_c) is (1921.5.3.1):

$$\phi V_c = \phi 20 \sqrt{f_c'} A_j$$

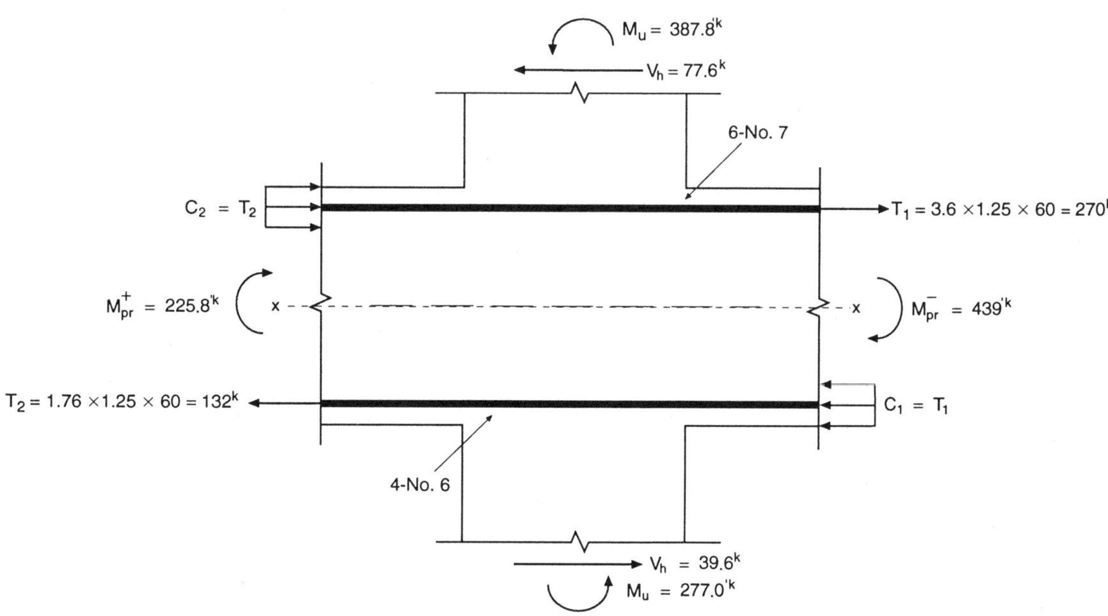

Figure 6-17 Shear Analysis of Interior Beam-Column Joint Along Column Line 6

where A_j is equal to the effective cross-sectional area within a joint in a plane parallel to the plane of the reinforcement generating shear in the joint. The joint depth is the overall depth of the column. The effective width of the joint shall not exceed the smaller of (1) the beam width plus the joint depth = 20 + 26 = 46 in. or (2) twice the smaller perpendicular distance from the longitudinal axis of the beam to the column side = 2 × [(20/2) + 3] = 26 in. (governs). Thus, A_j = 26 × 26 = 676 in.^{2} and:

$$\phi V_c = 0.85 \times 20 \times \sqrt{4{,}000} \times 676 / 1{,}000 = 726.8 \text{ kips} > 362.4 \text{ kips} \qquad \text{O.K.}$$

Exterior Joint.

The shear strength will be checked in the N-S direction for the exterior beam connection along column line 6. The edge column has beams framing on three sides; thus, the joint is not confined by framing members according to 1921.5.2.2 so that the transverse reinforcement required for the column must be continued through the joint.

Figure 6-18 shows a section through an exterior beam-column joint at the first floor level. The shear force at section *x-x* is obtained by subtracting the column horizontal shear force from the tensile force in the top beam reinforcement.

The column horizontal shear force (V_h) can be obtained by assuming that the beam in the adjoining floor is deformed so that a plastic hinge form at the junction with the column, with M_{pr}^{-}(beam) = 302.6 ft-kips.

It is assumed that the beam moment is resisted by the columns above and below the joint inversely proportional to their lengths. Thus, above the joint, the moment is equal to 302.6 × [14/(14 + 10)] = 176.5 ft-kips; below the joint the moment is equal to 126.1 ft-kips. The horizontal shear force at the top of the joint is then equal to:

$$V_h \text{ (top)} = \frac{2 \times 176.5}{10} = 35.3 \text{ kips}$$

and at the bottom of the joint it is:

$$V_h \text{ (bottom)} = \frac{2 \times 126.1}{14} = 18.0 \text{ kips}$$

The tensile force in the top reinforcement (4-No. 7 bars) is equal to (4 × 0.60) × 1.25 × 60 = 180.0 kips.

The net top shear at section *x-x* is T_1 - V_h (top) = 180.0 - 35.3 = 144.7 kips.

The net bottom shear at section *x-x* is T_1 - V_h (bottom) = 180.0 - 18.0 = 162.0 kips.

For a joint confined on all three faces, the nominal shear strength (ϕV_c) is (1921.5.3.1):

$$\phi V_c = \phi 15 \sqrt{f_c'} A_j$$

In this case, the effective width of the joint shall not exceed the smaller of (1) the beam width plus the joint depth = 20 + 22 = 44 in. or (2) twice the smaller perpendicular distance from the longitudinal axis of the beam to the column side = 2 × (20/2) = 20 in. (governs). Thus, A_j = 22 × 20 = 440 in.^{2} and:

$$\phi V_c = 0.85 \times 15 \times \sqrt{4{,}000} \times 440 / 1{,}000 = 354.8 \text{ kips} > 162.0 \text{ kips} \quad \text{O.K.}$$

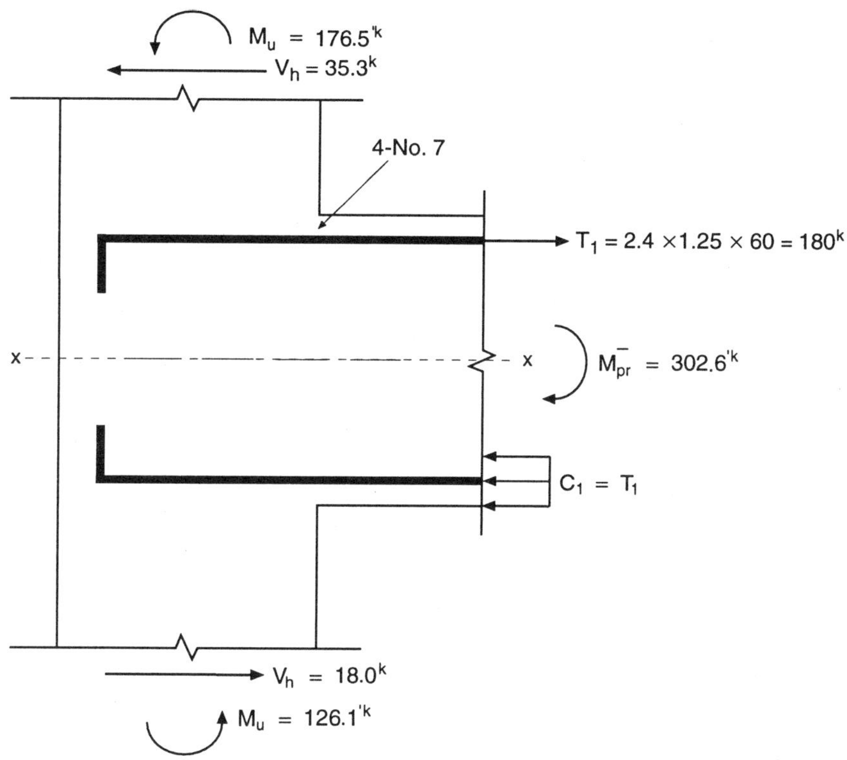

Figure 6-18 Shear Analysis of Exterior Beam-Column Joint Along Column Line 6

6.6.5 Shearwall Design

The design procedure for shearwalls can be summarized in three steps:

1. Design the shearwall for flexure and axial load by considering the entire cross-section of the wall to act as a short column (1921.6.6.1); this will yield the required flexural reinforcement. Check the shear resistance of the wall (1921.6.5).

2. Provide the boundary zone detail requirements of 1921.6.6.6 at each end of the wall unless $P_u \leq 0.10 A_g f'_c$ (for geometrically symmetrical walls) and either $M_u / V_u \ell_w \leq 1.0$ or $V_u \leq 3A_{cv}\sqrt{f'_c}$ and $M_u / V_u \ell_w \leq 3.0$, where ℓ_w is the length of the entire wall in the direction of the shear force and A_{cv} is the net area of the concrete section bounded by the web thickness and the length of the section in the direction of the shear force (1921.6.6.4). In any case, shearwalls with $P_u > 0.35 P_o$ are not permitted to resist earthquake-induced forces, where P_o is the nominal axial load strength of the wall at zero eccentricity (1921.6.6.3).

3. For shearwalls where boundary elements with specially detailed reinforcement have to be provided, determine the length of the boundary zone by using one of two methods. In the conservative method, the boundary zones are provided over a distance of $0.25\ell_w$ to $0.15\ell_w$ at each end of the wall (1921.6.6.4). Alternatively, the boundary zone length may be based on the compressive strain levels at the edges of the wall when the wall is subjected to displacements resulting from the ground motions specified in 1629.2 using cracked section properties of the wall (1921.6.6.5). Boundary zone detail requirements of 1921.6.6.6 must be provided wherever the compressive strain exceeds 0.003.

Combined Load Effects.

A summary of the design forces at the base of the shearwall along column line 7 is contained in Table 6-23 (see Tables 6-4 and 6-14).

Table 6-23 Summary of Factored Axial Loads, Bending Moments, and Shear Forces at the Base of the Shearwall Along Column Line 7

No.	Load Case	Axial Load (kips)	Bending Moment (ft-kips)	Shear Force (kips)
	Dead (D)	2,856	0	0
	Live (L)	265	0	0
	Earthquake (E_h)	0	49,005	774
	Wind (W)	0	15,933	306
No.	Load Combination			
1	$1.4D + 1.7L$	4,449	0	0
2	$0.75(1.4D + 1.7L + 1.7W)$	3,337	20,315	390
3	$0.9D + 1.3W$	2,570	20,713	398
4	$1.54D + 0.55L + 1.43E_h$	4,544	70,077	1,107
5	$0.77D + 1.43E_h$	2,199	70,077	1,107

Shear Design.

At least two curtains of reinforcement must be used in the wall between the boundary elements if the in-plane shear force (V_u) exceeds $2A_{cv}\sqrt{f_c'}$ where A_{cv} is the net area bounded by the web thickness and the length of the section in the direction of analysis (1921.6.2.2):

$$V_u = 1{,}107 \text{ kips} > 2A_{cv}\sqrt{f_c'} = 2 \times (18 \times 24.67 \times 12) \times \sqrt{4{,}000} \;/\; 1{,}000 = 674 \text{ kips}$$

Thus, two curtains of reinforcement are required in the web. Note that the reinforcement ratio along the longitudinal and transverse directions must not be less than 0.0025, and that the spacing of the bars must not exceed 18 in. (1921.6.2.1).

The upper limit on the shear strength of a shearwall is $\phi 8A_{cv}\sqrt{f_c'}$ which is equal to $0.6 \times 4 \times 674 =$ 1,618 kips (1921.6.5.6). Typically, the shear strength reduction factor is 0.85. However, this factor must be taken as 0.6 for any structural member if its nominal shear strength is less than the shear corresponding to the development of the nominal flexural strength of the member (1909.3.4.1). In this case, the more conservative value of 0.6 was used in the design of the wall; it is clear that the factored shear force is less than the limiting value.

Since the ratio of the height of the wall (h_w) to the length of the wall (ℓ_w) is equal to 148/24.67 = 6.0 the design shear strength of the wall (ϕV_n) is given by the following (1921.6.5.2):

$$\phi V_n = \phi A_{cv}(2\sqrt{f_c'} + \rho_n f_y)$$

where ρ_n is equal to the ratio of distributed shear reinforcement on a plane perpendicular to the plane of A_{cv}.

Using 2-No. 5 bars at 8 in. on center results in the following:

$$\rho_n = \frac{2 \times 0.31}{18 \times 8} = 0.0043$$

$$\phi V_n = 0.6 \times (18 \times 24.67 \times 12) \times \left[2\sqrt{4,000} + (0.0043 \times 60,000)\right]/1,000 = 1,230 \text{ kips} > 1,107 \text{ kips O.K.}$$

The vertical reinforcement ratio (ρ_v) shall not be less than the horizontal reinforcement ratio (ρ_n) when the ratio h_w/ℓ_w is less than 2.0 (1921.6.5.5). Since h_w/ℓ_w is equal to 6.0, the minimum reinforcement ratio will be used. The required area of vertical steel in the web is equal to 0.0025 × 18 × 12 = 0.54 in.^{2}/ft. Using 2 layers of No. 5 bars, the required spacing is (0.62 × 12/0.54) = 13.8 in. Use No. 5 vertical bars on each face at 12 in. on center.

Design of Boundary Zone.

Determine if boundary zone details are required. From Table 6-23, the maximum factored axial load (P_u) is equal to 4,544 kips. Check the criterion of 1921.6.6.4:

$$0.10 A_g f_c' = 0.10 \times [(18 \times 19.33 \times 12) + (2 \times 32^2)] \times 4.0 = 2,490 \text{ kips} < P_u = 4,544 \text{ kips}$$

Therefore, the boundary zone details of 1921.6.6.6 are required. Assume that the 32 × 32 in. boundary elements are reinforced with 32-No. 11 bars. Also, it was determined above that 2-No. 5 bars at 12 in. spacing is required as vertical reinforcement in the web. Based on this reinforcement, the nominal axial load capacity of the wall at zero eccentricity (P_o) is:

$$P_o = 0.85 f_c'(A_g - A_{st}) + f_y A_{st}$$

$$= [0.85 \times 4 \times (6,224 - 112.24)] + (60 \times 112.24) = 27,514 \text{ kips}$$

Since P_u is equal to 4,544 kips which is $0.17P_o$, conservatively provide boundary zones at each end a distance of 0.16 ℓ_w = 0.16 × 24.67 = 4.0 ft (1921.6.6.4).

Alternatively, determine the length of the boundary zone using the provisions of 1921.6.6.5. Before proceeding with the strain compatibility analysis, determine if the wall can be considered to contribute to the strength of the structure for resisting earthquake-induced forces. Since P_u is equal to 4,544 kips which is less than $0.35P_o$ = 9,630 kips, the wall can be used to resist its portion of the seismic loads.

When performing the strain compatibility analysis for the wall, it is important to note that the compressive strain in the concrete is not allowed to be larger than 0.015 (see Eq. (21-8)).

For shearwalls in which the flexural limit state is governed by yielding at the base of the wall, the total curvature demand (ϕ_t) can be determined from Eq. (21-9):

$$\phi_t = \frac{\Delta_i}{\left(h_w - \ell_p/2\right)\ell_p} + \phi_y$$

where $\Delta_i =$ inelastic deflection at the top of the wall = Δ_t - Δ_y

$\Delta_t =$ total deflection at the top of the wall equal to Δ_M, using cracked section properties, or may be taken as $2\Delta_M$, using gross section properties

$\Delta_M =$ maximum inelastic response displacement = $0.7R\Delta_S$

$\Delta_S =$ design level response displacement

$\Delta_y =$ displacement at top of wall corresponding to yielding of the tension reinforcement at the critical section, or may be taken as $(M'_n / M_E)\Delta_E$, where M_E is equal to the unfactored moment at the critical section when the top of the wall is displaced Δ_E and M'_n is the nominal flexural strength of the critical section at an axial load $P'_u = 1.2D + 0.5L + E$

$\Delta_E =$ elastic design displacement at the top of the wall using gross section properties and code specified seismic forces

$\ell_p =$ height of the plastic hinge above the critical section, which may be taken as $0.5\ell_w$

$\phi_y =$ yield curvature, which may be estimated as $0.003/\ell_w$

Note that if ϕ_t is less than or equal to $0.003/c'_u$, where c'_u is the neutral axis depth at P'_u and M'_n, boundary zone details are not required. If ϕ_t exceeds $0.003/c'_u$, the compressive strains may be assumed to vary linearly over the depth c'_u and have a maximum value equal to the product of c'_u and ϕ_t.

The first step in the analysis is to determine the location of the neutral axis c'_u. Using Table 6-23, the load P'_u is equal to (1.2 × 2,856) + (0.5 × 265) + 0 = 3,560 kips. The nominal moment strength (M'_n) corresponding to P'_u is 110,729 ft-kips (see Fig. 6-19).

In lieu of hand computations, *PCACOL* (Ref. 6.3) was used to obtain the neutral axis depth c'_u corresponding to P'_u and M'_n. For 32 × 32 in. boundary elements reinforced with 32-No. 11 bars and an 18 in. web reinforced with 2-No. 5 bars at 12 in., c'_u is equal to 58.7 in.

Figure 6-20 shows the interaction diagram with the load combinations that are contained in Table 6-23. As can be seen from the figure, the wall is adequate to carry the factored loads.

The structure was reanalyzed using the gross section properties of the members and the code-specified seismic forces in the N-S direction. From this analysis, the elastic displacement at the top of the wall (Δ_E) is equal to 1.32 in. and the corresponding bending moment (M_E) at the base of the wall is 43,376 ft-kips. From the analysis using the cracked section properties, the total displacement at the top of the wall (Δ_t) is equal to Δ_M = 14.2 in. (see Table 6-5). Also,

$$\Delta_y = \frac{M'_n}{M_E}\Delta_E = \frac{110{,}729}{43{,}376} \times 1.32 = 3.4 \text{ in.}$$

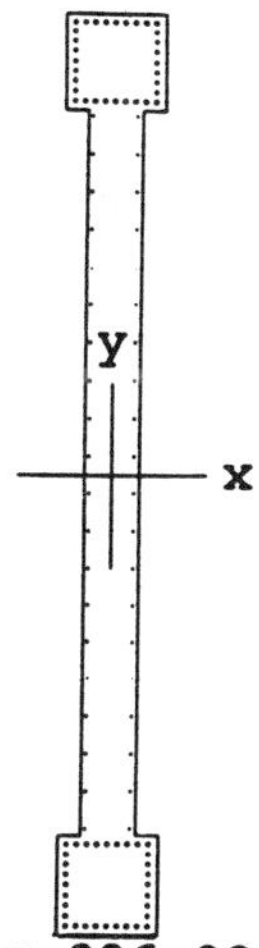

32.00 x 296.00 in

104 bars at 1.80%
As = 112.24 in^2
Cover = N/A
Spacing = 1.79 in

Ag = 6224 in^2
Ix = 54589864 in^4
Iy = 287515 in^4

Xo = 0.00 in
Yo = 0.00 in

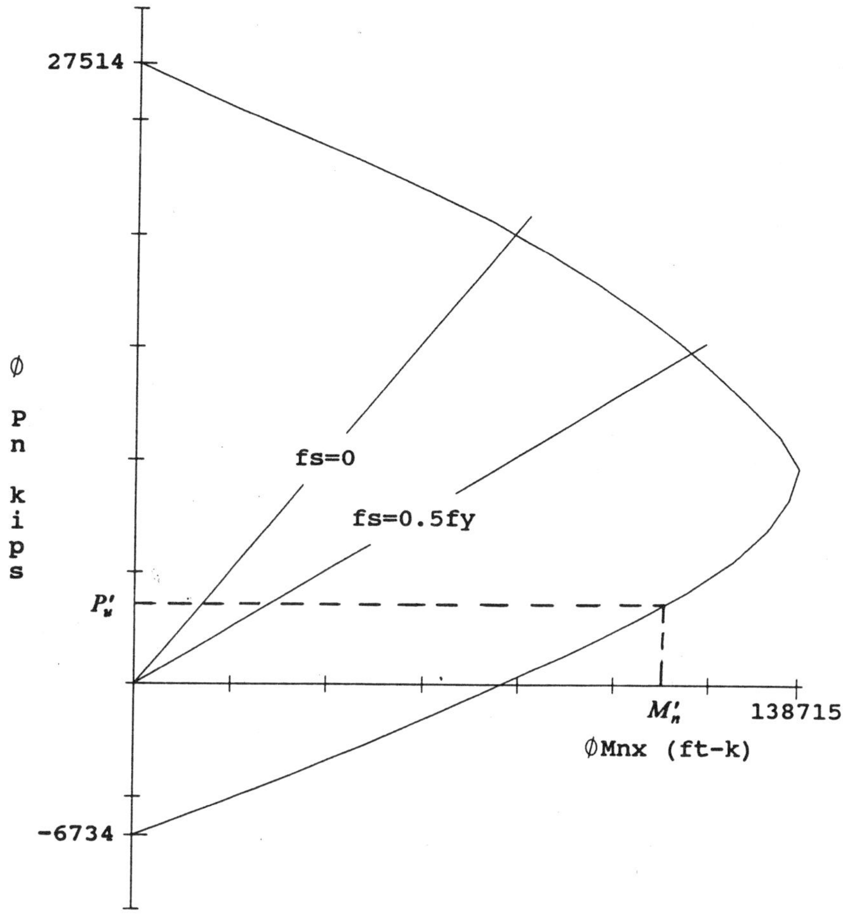

File name: C:\DATA\COL\UBC97WAL.COL

Project: UBC-97	f'c = 4.00 ksi Ec = 3834 ksi
Column Id: Shearwall - Line 7	fc = 3.40 ksi eu = 0.003 in/in
Engineer:	Block stress profile, Beta1 = 0.85
Date: Time:	fy = 60 ksi Es = 29000 ksi
Code:	Confinement: Other
Units: English	phi(c) = 1.00, phi(b) = 1.00, a = 1.00

X-axis slenderness is not considered.

Figure 6-19 $P_n - M_n$ Diagram for the Shearwall on Column Line 7

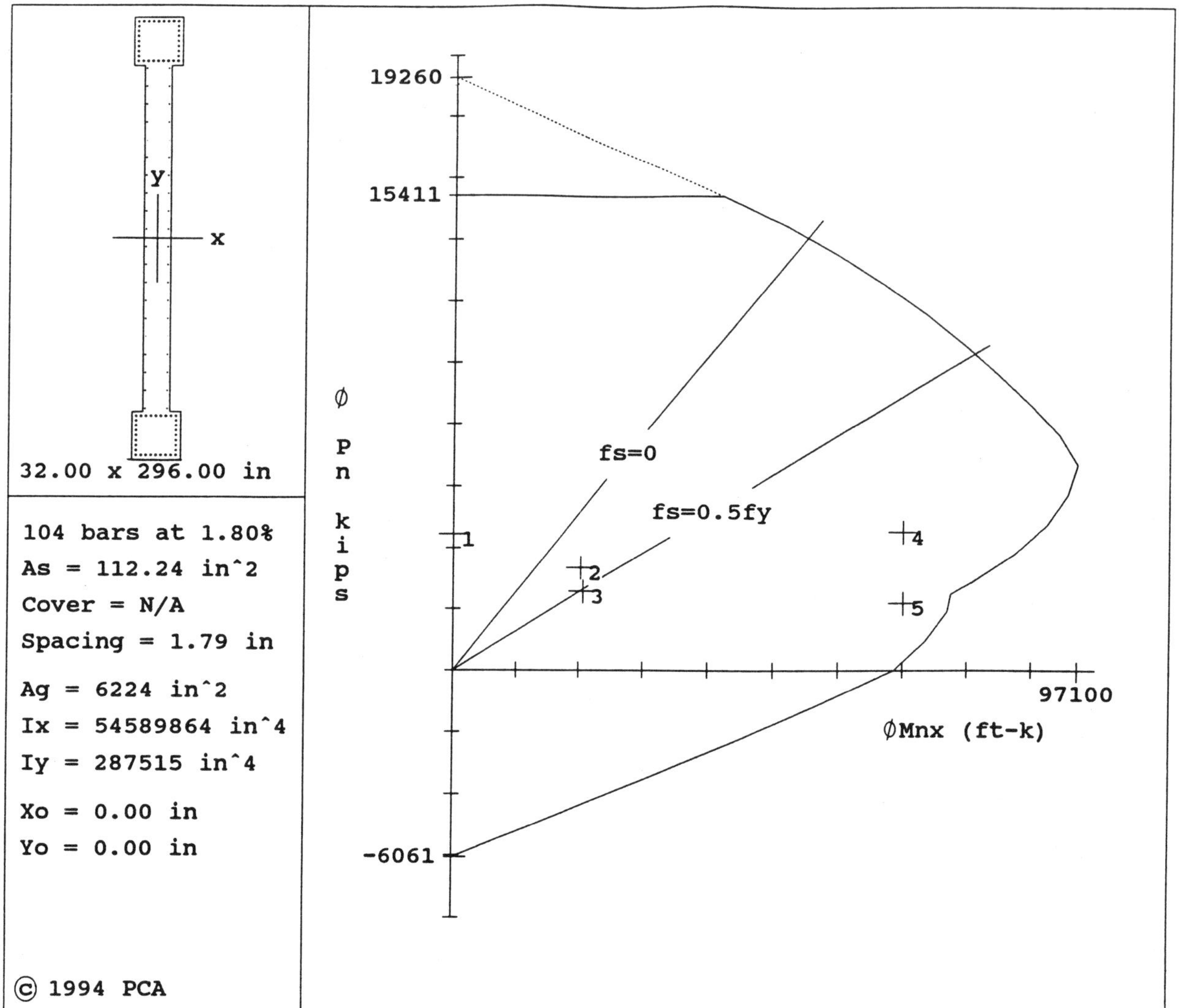

Licensed To: Portland Cement Association, Skokie, IL

File name: C:\DATA\COL\UBC97WAL.COL

Project: UBC-97	f'c = 4.00 ksi Ec = 3834 ksi
Column Id: Shearwall - Line 7	fc = 3.40 ksi eu = 0.003 in/in
Engineer:	Block stress profile, Beta1 = 0.85
Date: Time:	fy = 60 ksi Es = 29000 ksi
Code:	Confinement: Tied
Units: English	phi(c) = 0.70, phi(b) = 0.90, a = 0.80

X-axis slenderness is not considered.

Figure 6-20 Design Strength Interaction Diagram for the Shearwall on Column Line 7

The inelastic deflection at the top of the wall is:

$$\Delta_i = \Delta_t - \Delta_y = 14.2 - 3.4 = 10.8 \text{ in.}$$

Assuming that ℓ_p is equal to $0.5\ell_w = 0.5 \times 24.67 \times 12 = 148$ in., the total curvature demand is:

$$\phi_t = \frac{10.8}{[(148\times 12)-(148/2)]\times 148} + \frac{0.003}{24.67\times 12} = 4.29\times 10^{-5} + 1.01\times 10^{-5} = 5.30\times 10^{-5}$$

Since ϕ_t is greater than $0.003/c'_u = 0.003/58.7 = 5.11 \times 10^{-5}$, boundary zone details as defined in 1921.6.6.6 are required. Also, the compressive strains may be assumed to vary linearly over the depth c'_u with the maximum compressive strain equal to $\phi_t c'_u = 5.30 \times 10^{-5} \times 58.7 = 0.0031$ which is less than ε_{max} which is 0.015. Confinement is required over

$$\left(58.7 - \frac{0.0030}{0.0031}\times 58.7\right) = 1.9 \text{ in.} < 32 \text{ in.}$$

For practical purposes, the entire 32 × 32 in. boundary element will be confined according to the provisions of 1921.6.6.6. Check minimum dimensions of the boundary zone:

$$\text{Minimum thickness} = \frac{\ell_u}{16} = \frac{(16\times 12)-24}{16} = 10.5 \text{ in.} < 32 \text{ in.} \quad \text{O.K.}$$

Minimum length = 18 in. < 32 in. O.K.

The minimum area of confinement reinforcement (A_{sh}) is given in Eq. (21-10):

$$A_{sh} = \frac{0.09 s h_c f'_c}{f_{yh}}$$

Using the maximum allowable spacing of $6d_b$ = 6 3 1.41 = 8.5 in. or 6 in. (governs) and No. 5 hoops:

$$A_{sh} = \frac{0.09\times 6\times[32-(2\times 1.5)-0.625]\times 4}{60} = 1.02 \text{ in.}^2$$

No. 5 hoops with crossties around every other longitudinal bar provide $A_{sh} = 5 \times 0.31 = 1.55$ in.^{2} Note that over the splice length of the vertical bars in the boundary zone, the hoops and crossties must be spaced at 4 in. on center. Also, the minimum area of vertical reinforcement in the boundary zone is $0.005 \times 32^2 = 5.12$ in.^{2} which is less than the area provided by 32-No. 11 bars. Figure 6-21 shows the reinforcement details for the shearwall.

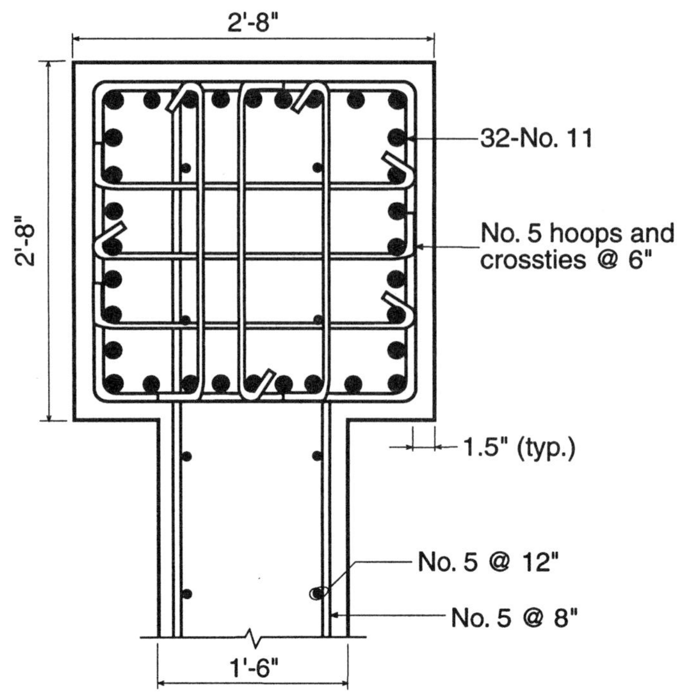

Figure 6-21 Reinforcement Details for the Shearwall on Column Line 7

6.7 References

6.1 *SAP-90 — A Series of Computer Programs for the Finite Element Analysis of Structures,* Computers and Structures, Inc., Berkeley, CA, 1992.

6.2 *Notes on ACI 318-95, Building Code Requirements for Structural Concrete with Design Applications*, Portland Cement Association, Skokie, IL, 1996.

6.3 *PCACOL — Strength Design of Reinforced Concrete Column Sections,* Portland Cement Association, Skokie, IL, 1994.